Inertia *IS* Gravity

Masses, Forces, Motions, and a Number a Little Bit Bigger Than

137

An
Essay
by

Guy Cavet Myhre

authorHOUSE™

1663 LIBERTY DRIVE, SUITE 200
BLOOMINGTON, INDIANA 47403
(800) 839-8640
WWW.AUTHORHOUSE.COM

First published by AuthorHouse 10/07/04

ISBN: 1-4208-0712-9 (sc)

Printed in the United States of America
Bloomington, Indiana

This book is printed on acid-free paper.

Dedicated to my beloved wife, Annick.

Table of Contents

Preface

In this essay, I present previously unknown information about our physical environment. I begin with explanations about characteristics of forces that control every aspect of our human lives, of the lives of every living being in the Universe, and also of rocks, mountains, planets, stars, galaxies, and so on. The magnitudes of these forces establish the size of atoms in our bodies and of all the objects in our environment. They also control our rate of aging and the rate of creation, growth, decay, and rotting of everything around us. These force magnitudes determine the rapidity of our reasoning, of our thinking, and of our physical movements—the time required to blink an eye, to lick a lip, to take a step, to wave an arm, to come to a decision, to trigger a reflex action, or to accelerate a dragster.

The magnitudes of these forces appear to vary depending upon the location in the Universe of the application of these forces. A particular density of massive objects—stars and galaxies—surround each location. Some locations possess low-mass density and some, high. In low-mass-density space, things are relatively small and their progressions happen fast. In higher-mass-density space, these *same* things are bigger and their progressions happen slower. In this essay, we discover that the value of, what I call, the *mass-density index* is, in the vicinity of our Solar System, probably *a little bit bigger than 137.* A few million years into the future, when our Solar System will be nearer the Black Hole at the center of our Milky Way galaxy, that index number will probably increase. Millions of years in the past, it was certainly less.

The mass-density index possesses the same value as does the *reciprocal of the fine-structure constant* (α^{-1}), which constant Arnold Sommerfeld introduced in 1916 while studying the *fine structure* of spectral lines in atoms. It never was thought to have anything to do with the relative magnitudes of inertial forces between various locations in the Universe, but now I believe that it does. Those all-powerful, pervasive controlling forces are historically called *inertial forces*, which, now, can be called, more precisely, *aggregate gravity*—**Inertia *IS* Gravity**.

While studying the awesome aspects of the ultra-powerful inertial forces and manipulating the historical force equations, I discovered that their force constants, which are considered to be fundamental, are *definitely* not so. Each force constant is composed of a particular combination of mathematical factors, which is an assortment of unidimensional quantum attributes of the elementary particle to which they pertain, *except* for one—the *fine-structure-constant* factor.

Discovering that the famous dimensionless fine-structure constant is a factor in *all three* of the force constants—electric, magnetic, *and gravitational*—excited me no end. From there, everything else of pertinence to my range of interest seemed to fall into place. I discovered the simple, yet elegant, mathematical linkage between electromagnetic and gravitational force as well as the linkage between the unidimensional quantum attributes of the electron and proton, *and* between the electron and *masstron* (sic).

Then, extending this new discovery process to all of the, so called, *fundamental universal physical constants of nature*, I found that most of them are not fundamental at all. Each one of them is like the force constants—composed of mathematical factors, which are the unidimensional quantum attributes of the elementary particle to which the constant relates. Also, by creating and using a system of unit measures that employs these unidimensional quantum attributes as its units, most of these constants each acquire a value of *one*; therefore, as mathematical factors in equations, they can be removed from the equations without invalidating them. In essence, these fundamental universal physical constants of nature *do not exist in nature.* They exist only because they needed to be *invented* (not discovered) and placed in equations that use the metric system of unit measures, which units possess humongous magnitudes relative to those of elementary particles. As we will learn in this essay, this makes the venerable metric system of unit measures wholly inadequate for use in equations that relate to elementary particles. The fine-structure constant (a local constant), along with the speed of light (a universal constant), was *discovered* (not invented).

Reading the content of this essay will clarify all of the information that this preface summarizes. Although this content is targeted to laïcal physicists like me, professional physicists will find that it revolutionizes hypotheses concerning inertial force, topples the current beliefs about *fundamental universal physical constants of nature*, and rationalizes the basis for formulating simpler equations that relate better to quantized phenomena. Further, these academic physicists, with their extended intuition, experience, and mathematical mastery, can take my basic discoveries and use them, better than I, to realize new heights of knowledge of our physical environment.

Guy Cavet Myhre
San Diego, California
August 2004

Part One:

The Voyage

I. INTRODUCTION

I invite you to accompany me on a fantastic voyage—a voyage of discovery traversing the Universe—or, at least, our Milky Way galaxy. First, we must prepare ourselves for the hardships of the trip by transforming ourselves into super humans who have need for neither air, heat, cold, food, water, nor other life-sustaining resources—but have the ability, under the influence of various forces, to move about and perform our task of discovering the realities of our environment. Of course, this voyage is an imaginary one, which occurs only in our minds—in our thoughts. Yet, it is an effective tool for increasing our ability to understand the reality of the space into which we all are born. To succeed in charting a succinct voyage for us, I needed to undertake it alone beforehand so that, during our trips together, we will be better able to avoid the pitfalls that lurk along the way.

My own voyage began over a half-century ago when I wondered from where came the mysterious force that tried to run my car off the road in a sharp curve. Later, as a USAF pilot, I suffered this same weird force over and over again. This strange force, of course, is inertial force, which possesses magnitudes similar to those of gravitational force. Yet, inertial force can be *very* much more powerful than the gravitational force of Earth—to wit, a pilot-training centrifuge gone wild can *kill* people while they are sitting down. From where does this killing force come? Searching university physics libraries for information about inertial force garnered precious little. Curiously, I did find that inertial force is considered to be an artificial or pseudo force, which would shock many people who are *dead* because of the morbid reality of it—yes, inertial force is *very* real.

As the decades passed, my interest in the enigma of inertial force increased until thinking about it and about related physical phenomena became a hobby. As spare time permitted, I studied many books about astronomy, cosmology, and quantum physics without understanding much of the arcane mathematics, but I slowly began to understand *some* of the work of geniuses from Copernicus to Wheeler. This understanding armed me with a basic knowledge that enabled me to ponder intelligently about those phenomena of a quizzical or, perhaps, paradoxical nature that most interested me—masses, motions, velocities, accelerations, *and* the three extranuclear forces: electromagnetic, inertial, and gravitational. Eventually, the result of my pondering and calculating seemed to indicate that inertial force is actually *aggregated* gravitational force on a grand, universal scale—the application, at *any* particular location in the Universe, of *all* gravitational forces in the Universe, which are applied at that location—pulling electrons, which are bound to atomic nuclei, outward in every three-dimensional direction.

Then, upon examining equations about quantized phenomena concerning elementary particles, I noticed that they contain a profusion of what are called *fundamental universal physical constants of nature* (see Tables VI and VII). This is an important-sounding, yet confusing, name, which, in reality, should be called simply *constants of proportionality*. Many of these "constants" possess extremely small numerical values, which are that way because the metric units of measure used in them possess, in comparison, extremely big magnitudes. After much juggling of figures, I invented a new *Système électronique* (SE) of unit measures (see Section II.H. in *Part Two: The Calculations*) that is based upon, what I call, the *unidimensional quantum attributes of the electron*. Using this new *quantum* system of unit measures in equations that pertain to the electron amazed me—*the constants vanish*—that is to say, they acquire values of unity (see Table VI. under the heading of the column labeled **SE value**) and, as arithmetic *factors*, can be removed from the equations without invalidating them. Without the presence of encumbering constants in equations, their meanings are much easier to understand, which made discovering heretofore hidden truths about physical phenomena much easier to realize.

Yet, a problem persisted: When using the new units, the three *force* constants—for electrical, magnetic, and gravitational forces—do *not* possess values of unity; they each possess a *number a little bit bigger than 137* or, to be more exact, 137.03599911(46), which is the *current* value of the reciprocal of Arnold Sommerfeld's fine-structure constant (α^{-1}) as listed by the *National Institute of Standards and Technology*, accessible at http://physics.nist.gov/cuu/constants/index.html.

Nobody seems to agree as to the significance of the dimensionless number 137. Even Richard Feynman, himself, once stated that "physicists put this number up on their wall and worry about it." Now, if you are unsure as to the importance of the fine-structure-constant enigma, key it into Google search and become awestruck when *over* 70,000 sites become available, each pertaining to the irritatingly-pervasive fine-structure constant. I racked my brain for a long time trying to discover why the force constants, in *particular*, contain the fine-structure constant as a mathematical factor. I drew a blank at physics libraries because the literature there does not acknowledge that the force constants contain the fine-structure constant or even are composed of other more-basic factors. This, in examining *Part Two: The Calculations*, we will find to be true.

I often awake in the middle of the night and, in the tranquility of silence and darkness, think about the problems encountered during the previous day. This particular moment in time enables me to create and stimulate my most-productive thinking (and also enables most other persons to do so too, I suspect). One night, the most-probable answer to the force-constant problem came to me in a flash—possibly, the fine-structure constant exists in force constants because, historically, the electrical, magnetic, and gravitational forces were measured using the inertial-force-based unit of force—the metric *newton*. Had either an electrical-, magnetic-, or gravitational-based unit of force been used to arrive at the value for its related force constant, that value, most probably, would have *excluded* the fine-structure constant. This seems to mean that *the fine-structure constant is closely associated with inertial force.*

Later in this essay, in *Part Two: The Calculations*, we discover that the more-plausible definition of the fine-structure constant is that it is the ratio between two magnitudes of inertial force. The numerator could *possibly* represent the magnitude of inertial force in an empty (hypothetical) universe. Then, the denominator would *possibly* represent the magnitude of inertial force in the vicinity of the Solar System. Therefore, we could say that, *probably*, inertial force in the Solar System is about 137 times stronger than that in a universe otherwise devoid of matter—an incredible concept with mind-boggling implications, which we explore during our voyage of discovery. Yet, let us not be tempted to summarily dismiss this fantastic concept as being ludicrous before we have finished our voyage together and gone on to examine the heart of this essay—the content of *Part Two: The Calculations.*

The current definition of the fine-structure constant α, using *le Système international* (SI) of metric unit measures, is $\alpha = 2 \pi e^2 (h c)^{-1}$. I discovered that this is *meaningless* because, when we factorize the three "defining" constants, (e^2, h, and c), into their more-basic constituents and cross cancel matching factors, only α remains, so $\alpha = \alpha$ (see Eq. <u>67</u> in *Part Two: the Calculations*) ; therefore, those three "defining" constants do *not* define the fine-structure constant, which is more basic than they. This e^2, h, and c combination of factors in the fine-structure constant has, in the past, prompted some people to believe that it pulls the disciplines of electromagnetics, relativity, and quantum physics together into one constant. The content of *Part Two: The Calculations* shows that this concept is *false.*

Apparently, the magnitude of the fine-structure constant, as a local (systemic) constant, cannot be calculated by discovering a *purely*-mathematical formula, which has been tried in vain innumerable times by many hopeful, yet unsuccessful, quasi-numerologists. It is like trying to find a *purely*-mathematical formula for the magnitude of g, the acceleration of gravity at the surface of Earth, which is also a local constant. Of course, by knowing the magnitudes of the mass and radius of Earth and Newton's gravitational constant, *big G*, we can *calculate* the magnitude of *little g*.

II. DEFINITIONS

While we travel and experience celestial phenomena together, we encounter many confusing terms, expressions, and key words. To understand each other during our voyage together, we need to agree as to their definitions, which this section explains. Perhaps you may consider skipping this section and going directly to Section III. I suggest that you not do so because I use new definitions that never have had to be used before. Definitions appropriate for *Part Two: The Calculations* are explained at the beginning of that section of this essay. Defined terms are **bolded** upon their first encounter.

A. Object, free-fall, attribute, elementary particle, quantum,mass

For the purpose of this essay, an **object** is a being, body, particle, or entity of self-contained existence within definite boundaries. It possesses mass, which creates a magnitude of gravitational force that is proportional to the magnitude of the object's mass and radiates outward in all three-dimensional directions, colliding with and attracting other objects. Examples of objects are: elementary particles, atoms, molecules, cells, bacteria, viruses, rocks, iguanas, books, tables, chairs, cars, airplanes, people, fish, warthogs, planets, moons, stars, galaxies, clusters of galaxies, et cetera.

A **free object** has only gravitational, inertial, or both of these forces applied to it. A free object is not subjected to by any external mechanical forces. An object in **free-fall** is a free object under the influence of only gravitational, inertial, or both of these forces. No mechanical forces are acting upon it; therefore, it is "weightless"; it does not "feel" any external forces acting upon it.

An **attribute** is an inherent trait, quality, property, or characteristic of an object. Examples are: color, density, texture, shape, viscosity, hardness, malleability, length, width, depth, mass, charge, temperature, energy, spin, wavelength, lifetime, et cetera. An attribute can possess a magnitude and several values for that magnitude. For example, "The magnitude of the length of that scary lizard is as long as my arm." The possible values of that "long-as-my-arm" magnitude could be 2 feet or 24 inches or ⅔ yard or 0.6096 meters, or the values could even be expressed in the appropriate number of ångström units, light years, parsecs, et cetera.

An **elementary particle** is an object that appears not to be composed of more-fundamental objects. The prime example is the electron. The proton is considered to be one also, even through it, supposedly, is composed of more-elementary quarks. This is because, as yet, no one has either detected the disintegration of a proton into constituent quarks or devised a way of disintegrating it.

A **quantum** is the smallest possible magnitude of an attribute of an object. For example, the smallest possible magnitude of an electrical charge is the charge of an electron. The next increment of charge is that of two electrons. This is because the electron is an elementary particle, which cannot be divided into smaller (fractional) objects.

Mass is an attribute of an object, and it is an indication of the amount of atomic or molecular material that the object contains. The magnitude of the mass of an object is directly proportional to the magnitudes of the gravitational forces that other, external objects apply to that object.

B. Magnitude, value

In this essay, the two terms, **magnitude** and **value**, mean different aspects of an attribute of an object. Magnitude refers to the absolute physical size of an attribute of an object, regardless of units of measure. For example, the magnitude of a building's height remains the same; however, the value of the magnitude is a function of the unit of length that is applied, such as inches, feet, yards, meters, ångström units, or light years.

A constant can possess a magnitude. For example, the magnitude of Planck's constant (h) is the combination of the magnitudes of its constituent quantum factors (see Eqs. 2 and 3 in *Part Two: The Calculations*), which are *unidimensional quantum attributes of the electron*. Each attribute possesses only one magnitude, but that magnitude can be expressed by one of a multitude of values, depending upon the system of unit measures used.

C. Force vector

A **force vector** graphically represents the magnitude and direction of a force. It is a straight line with an arrow head at the end that points in the direction that the force is applied. The length of the line indicates the magnitude of the force. For example: a three-inch-long line might represent 12 newtons of force at a scale of one-quarter inch to one newton.

When adding force vectors, their directions must be taken into consideration. The following thought experiment clarifies the concept of vector addition: Let a group of wranglers in a corral surround a mustang with each of their lassos around its neck. The wranglers pull on the mustang from many directions, which immobilizes it. For each lasso that pulls in a particular direction, another (conjugate) lasso pulls in the opposite direction with the same magnitude of force. Thus, the sum of the magnitudes of these two forces is zero. On a larger scale, all of the pulls of gravity from all of the objects in the Universe pull on the Solar System from all directions. Their *vectorial* sum accelerates it in a particular direction, yet the multitude of opposing force-vector conjugates hold the Solar System and all objects within it in a *tremendous* grip. This "inertial" grip rules our lives such that the magnitude of Earth's gravitational force appears to be minuscule in comparison, but, because most of the inertial forces "cancel out" each other, we are normally unaware of their existence. Yet, they manifest themselves most forcefully outside our immediate awareness, for example: gyroscopes, centrifuges, Foucault's pendulum, a pail of water whirled overhead, and the stability of the orientation of the Sun and of its satellites.

D. Velocity, acceleration

The **velocity** of an object is composed of speed and direction. Speed indicates the rate that an object is moving relative to another object. Direction indicates the "direction" in which the object is moving relative to that other object. Each object in the Universe possesses a great multitude of relative velocities but *no absolute* velocity. For example: a car on a road is assumed to have *only* one velocity—its forward, longitudinal velocity relative to the road, yet it has zero vertical and zero lateral velocities relative to the road, unless the driver zig zags or the road has pot holes. Also, the car has velocities relative to the train on tracks parallel to the road, to the hitchhiker walking beside the road, to the airplane streaking overhead, to the planet Mars, and to the center of the Milky Way's Black Hole, to name a few. An object's velocity relative to another object can change when either of the two objects is accelerated. In essence, when an object is accelerated, its velocities relative to all other objects are changed and vice versa.

An **acceleration** changes the velocity of an object relative to another object over a period of time. An external force that is applied to an object creates the acceleration. Like with velocity, every object possesses a different rate of acceleration relative to each of the other objects in the Universe. A first object's acceleration relative to a second object can change even though the first object has no new external forces applied to it. If the second object has external forces applied to it, the accelerations of both objects change relative to each other.

However, unlike with velocity, each object also possesses an *absolute* acceleration whose magnitude is proportional to the vector sum of all of the external forces—both gravitational and mechanical—from all of the other objects in the Universe, which are applied to that object. As a result, *every object in the Universe is accelerating absolutely.* An object may not necessarily be accelerating relative to another object, for example: airplanes flying in formation are not accelerating relative to each other; however, they *are* accelerating *absolutely*, even when they are *not* accelerating relative to the air through which they are flying.

E. Gravity line

A **gravity line** is the path in four-dimensional space-time that a free object follows. It can be considered to be a type of geodesic or world line but is more restrictive in its definition. A free object's gravity-line path is a "lazy" line, which means that the free object has no other external force applied to it other than the natural gravitational forces, which every object is subjected to. A gravity line is not necessarily a Euclidean straight line; it may wander left, right, up, or down as the free object travels through the gravitational fields of other free objects, which pull it toward them by their own gravitational forces as it passes.

F. Center of gravity, front, back, push, pull

The **center of gravity** of an object is also its center of mass. By dividing an object in "half" by inserting an imaginary flat plane, in any orientation, through the center of gravity of the object, half of the object's mass is on one side of the plane, and the other half is on the opposite side of the plane.

The **front** of a free object faces the direction of that object's absolute acceleration.

The **back** of a free object faces the direction from the force that accelerates it.

A **push** is an external force that accelerates a free object by applying itself in *back* of the free object's center of gravity. If this push is created by a mechanical force, it creates an unstable condition such that, if the vector that represents the push does not go *exactly* through the object's center of gravity, the object rotates or changes direction. In the late 1950s, when the United States was trying to catch up to the Soviet Union's success with Sputnik, the thrust of some of its early space vehicles' rockets could not be made to stay in line with the vehicles' centers of gravity, and the vehicles had to be destroyed shortly after liftoff when they began to spin out of control. Thankfully, gravitational and inertial forces do not push; they pull, but, in essence, they do neither because their forces are automatically applied at the center of mass of *each* elementary particle within and throughout the object—front-to-back and left-to-right. Most mechanical forces are "pushes."

A **pull** is an external force that accelerates a free object by applying itself in *front* of the free object's center of gravity. This creates a stable condition such that, if the vector that represents the pull does not go *exactly* through the object's center of gravity, the object will change direction but only to follow the source of the force. An ocean-going tug boat pulls its barge with little regard to whether the barge is going in the right direction—the barge follows in the wake of the tug boat. All gravitational and inertial forces are "pulls."

G. Forces

External **forces** pull or push a free object such that it accelerates or changes its multitude of different velocities relative to other objects. If no external forces are applied to an object, it does not (absolutely) accelerate. Forces that accelerate objects are of three types—gravitational, (what I call) mechanical, and inertial, which appears to be *aggregate* gravitational force—*Inertia IS Gravity.*

1. Gravitational force

An object, which possesses mass, creates **gravitational force** and attracts other objects over long distances. All objects gravitationally attract one another—elementary particle-by-elementary particle. According to Isaac Newton, the magnitude of the gravitational force of attraction between two objects is directly proportional to the product of the magnitudes of their two masses and inversely proportional to the square of the distance between their centers of gravity. A gravitational force does not apply itself to the exterior boundary of a solid object as does a mechanical force; it attracts on the elementary-particle level such that every elementary particle throughout the object is accelerated equally—every infinitesimally tiny particle is attracted in direct proportion to the magnitude of its particular mass. A gravitational force applied to a free object accelerates it, but the object "detects" no force—it is "weightless." All objects of different masses that are in a relatively-close vicinity of one another are accelerated with the same magnitude, but each one of those objects is acted upon by gravitational forces whose magnitudes are directly proportional to the magnitude of the mass of that object in the same manner as for the other objects.

We, as living, feeling beings on the surface of Earth, cannot "feel" gravitational force. The gravity of Earth tries to accelerate us toward the center of Earth, but the surface of Earth gets in the way. What we "feel" is not gravity; it is the mechanical force of the surface of Earth counteracting the pull of gravity—accelerating us "upward." We will understand these physical phenomena more clearly as we voyage through the cosmos together.

2. Mechanical force

An external **mechanical force** that pushes an object, accelerates it as does gravitational force, but mechanical force first applies itself only to the surface of the object that is facing the force. The magnitude of the force is relayed from the surface of the object into the interior, particle-by-particle, and through the object to its other side. The object "detects" mechanical force, and if the force is too strong, it may break up the object—destroy it. This does not occur with gravitational force. The prime example of mechanical force is objects colliding with each other—a fist in the face, airplanes crashing, cars hitting each other in an intersection, hot gasses expanding within a rocket, relative wind against skydivers, et cetera. An old, decrepit building does not collapse by the force of gravity; Earth pushing "up" on the building is the culprit. We have difficulty envisioning this phenomenon because we are born into and spend our entire lives within a "weighted" environment on Earth, which is *not* the normal "weightless" accelerating environment found throughout the Universe.

3. Inertial force

Any object at *any* particular location in the Universe has a multitude of gravitational forces from all of the other objects in the Universe—galaxies, stars, nebulae, planets, moons, comets, asteroids—pulling on it from all directions in three-dimensional space. Vectors that represent these gravitational forces radiate outward from that object like the spines of a sea urchin or the rays of the Sun. The *vectorial* sum of the magnitudes of all these vectors accelerates the object in a particular direction. Thus, every free object in the Universe is accelerating—is in free-fall. For example, the Sun is accelerating within the Orion spiral arm of the Milky Way galaxy and is carrying all of its satellites, including Earth, along with it.

Suppose that the vectorial sum of the magnitudes of all the gravitational forces that is applied to a particular (hypothetical) object is equal to zero.

Then, that object is not accelerating *absolutely*, yet, still applied to it are cancelling-out *gravitational* forces, which try to keep the object in its current state of motion. A mechanical force applied to that object would be resisted by the restraining surrounding *aggregated gravitational* forces—the *inertial* forces.

The *newton* is the SI metric unit for inertial force and, unfortunately, is also used for all types of force, as we will realize later. The *magnitude* of one newton was *defined* as the force necessary to be applied to a one kilogram free object for one second such that the object increases (or decreases) its velocity relative to other objects by one meter per second. We can imagine how powerful a newton is by visualizing it pushing, for one second, a one kilogram (2.2 pounds) mass at rest, resulting in that mass moving at a rate of one meter per second (2.23 miles per hour).

Another object at another location in the Milky Way galaxy, such as in the denser Sagittarius spiral arm of the Milky Way, has more-powerful surrounding, cancelling-out gravitational forces applied to it, in which case inertial forces at that location would be stronger than in the Orion spiral arm. For illustrative purposes, suppose that they are twice as strong; then, the *magnitude* of one newton would be twice as big, although its *value* would remain the same—at one. This means that the magnitude of inertial forces seem to vary throughout the Universe. Also, the magnitude of inertial forces, because they are, in essence, aggregate gravitational forces, appear to apply themselves on the atomic level and be directly proportional to the magnitudes of the distances from the nuclei of the atoms to the orbits of the electrons that are bound to those nuclei. This seems to mean that objects occupy more space in locations in the Universe that possess greater magnitudes of inertial forces. This appears to be a reasonable assumption because, after all, on the atomic level, inertial forces obviously try to pull the orbits of electrons, which are bound to atomic nuclei, away from those nuclei, as we will see in *Part Two: The Calculations*.

H. Linear motion

Gravitational and inertial forces applied to a free object linearly accelerate that object and do not cause it to rotate or change direction. This is so because gravitational and inertial forces do not apply themselves to the object as a whole but to each individual elementary particle within that object in direct proportion to that particle's mass; therefore, the forces balance themselves on either side of the object's center of gravity, which directs the object's linear motion along its gravity line. Of course, the gravity line, itself, may not necessarily be a Euclidean straight line if it should traverse a field of objects.

An external *mechanical* force applied to an object linearly accelerates that object provided that the force vector that represents that mechanical force is in line with the object's center of gravity. Otherwise, the mechanical force changes the object's direction or starts it to rotate. When an external mechanical force is applied to a free object, the object changes its acceleration, and an opposing inertial (gravitational) force, which had been cancelled out by an opposing inertial force of the same magnitude, suddenly manifests itself to counteract the mechanical force. Cancelling-out inertial forces applied to a free object in linear motion only manifest themselves in opposition to an external mechanical force when it is applied to that object.

I. Rotating motion

A free object is not necessarily an "unconstrained" object. When it is in **rotating motion**, it is constrained by internal forces—both electromagnetic and gravitational—which are the forces that bind together the elementary particles of which the free object is composed. From a distance, a rotating sphere appears not to move, yet every elementary particle of which that sphere is composed is continuously changing direction under the centripetal pull of internal electromagnetic and gravitational forces, which are opposed by the manifestation of, otherwise hidden, centrifugal inertial forces. However, in a homogeneous rotating sphere, every direction-changing elementary particle balances with its conjugate particle opposite the center of gravity and the same distance away from it. This phenomenon gyroscopically stabilizes the rotating motion of the sphere such that the orientation of its axis remains fixed relative to the orientations of all of the force vectors, which represent all of the pulls of gravity from all of the other objects in the Universe.

A car traveling down a straight road at a constant speed relative to the longitudinal axis of the road appears not to have forces applied to it (except for the mechanical force of Earth pushing up on it), yet undetected cancelling-out inertial forces are constantly being applied to it. Then, when the car enters a curve, the lateral centripetal mechanical force applied by the tires on the road invokes the opposing inertial centrifugal forces, which are always lurking—ready to strike.

J. Gravitons

Be forewarned: the information in this section is highly speculative and simplistic because we know little about gravitons, which really must exist, otherwise how could objects gravitationally attract each other over long distances. So, take my model for gravitational attraction with skepticism, but I hope that it will stimulate your own ideas so that you can create a more-plausible model, which I would be most anxious to peruse. Well, here it goes:

As far as we know (or believe we know), a **graviton** is a massless elementary entity that is emitted in a particular direction into three-dimensional space at the speed of light from an elementary particle, which can be called a *graviton-creating* particle. Apparently, a graviton follows a gravity line, capable of traveling over long distances, until it hits a second elementary particle, which, itself, emits streams of gravitons in like manner. When hit by the graviton from the first particle, the second particle is pulled toward the first particle by a *quantum unit of gravitational force*. Traveling at light speed, the magnitude of the gravitational force that a graviton creates does not change with lengthening of distance traveled—it remains at the fixed magnitude of the *quantum unit of gravitational force*.

1. Surface area of a sphere

Each object contains a multitude of elementary particles, each of which emits a multitude of graviton streams into three-dimensional space. The density of these graviton streams (or lines of force) declines as the distance from the object increases. With respect to various different-sized imaginary spheres concentric to the center of gravity of the object, each sphere is pierced over its entire surface by the same number of graviton streams; therefore, all points on the surface of a particular sphere are subject to the same density of graviton streams, whose directions are perpendicular to the surface of the sphere. This means that the magnitude of the gravitational force emanating from that object at a point on the surface of this sphere is inversely proportional to the magnitude of the sphere's surface area of $4 \pi r^2$, where r is the radius of the sphere.

2. Product of two masses

Let us suppose that two objects (O_1 and O_2) are a particular distance r from each other. Then, two imaginary overlapping spheres of radius r can be visualized such that each of them is concentric to the center of one of the objects and resides somewhere on the surface of the other object's sphere. Further, let us suppose that object O_1 contains m_1 numbers of graviton-creating particles, and O_2 contains m_2 of them. This means that the magnitudes of the masses of O_1 and O_2 are proportional, respectively, to m_1 and m_2. At any particular moment in time, each of the m_2 number of graviton-creating particles in object O_2 is hit by a graviton emitted from each of the m_1 number of graviton-creating particles in object O_1.

As a hypothetical microscopic example, let $m_1=3$ and $m_2=4$. Then, at any single moment in time, each of object O_1 's three graviton-creating particles emits four gravitons, each of which hits one of the four graviton-creating particles of object O_2 for a total of 12 graviton hits. Therefore, the magnitude of the gravitational force of attraction between objects O_1 and O_2 in this example are proportional to the 12 quantum units of gravitational force created from the three graviton-creating particles of object O_1. However, object O_2's four graviton-creating particles also emit a total of 12 gravitons that hit the three graviton-creating particles of object O_2 for a grand total of 24 quantum units of gravitational force.

In sum, the gravitational force of attraction between objects O_1 and O_2 is directly proportional to $2\ m_1\ m_2$ and inversely proportional to $4\ \pi\ r^2$. Of course, because 2 and $4\ \pi$ are constants, they could be removed from these results and placed into an eventual constant of proportionality when creating the equation from the proportion. This was done, of course, with Newton's gravitational constant G , but, by doing that, the physical meaning of what the Newton gravitational-force equation is supposed to indicate is obscured. This result is explained in more detail in *Part Two: The Calculations*.

Well, those are enough definitions for now; let us start our voyage of discovery together into deep space—into the cosmos.

III. VOYAGE INTO THE COSMOS

Our voyage into the cosmos consists of a series of trips, the first of which is to the center of Earth, and the last is to the inner cosmos of elementary particles, but fear not, we will be perfectly equipped to resist the dangers of the fierce environments into which we will be subjected because the voyage will exist in our minds alone. Before leaving the surface of Earth to descend into its bowels, we measure the magnitudes of three of our bodily attributes for later comparison with the magnitudes of these attributes when we are at other locations in the Universe.

On the surface of Earth, let each of our heights be two meters and our masses be 100 kilograms—so we are pretty hefty people. No reason for this large size exists except that these "easy" values make mathematical calculations less difficult. The third measurement is the more important one and the more difficult one to measure—the size of atoms in our bodies and, therefore, in every object in our environment. For the sake of simplicity, let us pick a simple atom—hydrogen—whose nucleus binds to itself only one electron at a particular minimum distance from itself, called the *Bohr radius of the hydrogen atom*. Reasonably, the magnitude of the Bohr radius is proportional to the radii of the orbits of all electrons, which are bound to the nuclei of all atoms in our environment. In sum, the magnitude of the Bohr radius is an indication of the "size" of all matter. In the vicinity of the Solar System, the magnitude of the Bohr radius r_B is about 137 times that of the electron Compton wavelength λ_e divided by 2π, which is $r_B = \lambda_e (2\pi\alpha)^{-1}$. The electron Compton wavelength is the quantum unit length in the SE of unit measures. See Table XI.

A. Center of Earth

At the surface of Earth, we enter an elevator that has a weight scale built into the floor. Never fear, this super elevator can descend the almost 4,000 miles to the center of Earth, keeping us in air-conditioned comfort. Before descending, we notice that all three types of extranuclear force are acting upon our bodies—*gravitational*, *inertial*, and *mechanical*.

The *first* type is the *vector* sum of all of the pulls of gravity from all of the surrounding objects in the Universe, which is accelerating us along with everything else in the Solar System in a particular spiral direction, apparently toward the Black Hole at the center of our Milky Way galaxy. The "downward" accelerating pull of Earth's gravity is included in this vector sum, although it is cancelled out by the mechanical force of Earth, which is accelerating us "upward."

The *second* type is all of those pulls of gravity from all of those surrounding objects in the Universe that oppose each other and cancel out each other so that they do not contribute to the vector sum but create the inertial forces instead.

The *third* type is the already-mentioned mechanical force of Earth accelerating us "upward."

The elevator begins our descent at a constant speed. Except for molten rock and increasing heat, we notice little change at first. Then slowly, the force of Earth's gravity appears to decrease because we "weigh" less and less as we continue to descend. This *unidirectional* pull of Earth's gravity is gradually converting to additional *omnidirectional* inertial forces, which results in our movements becoming slightly more sluggish; our speech, a little slower; our bodies, a bit bigger; and our aging process, a bit slower; however, we cannot notice these changes because our thought processes are slowing as well, and everything around us is getting bigger as we get bigger. In essence, the magnitudes of the quantum-length and -time units both increase by the same factor as we descend, such that the magnitude of their quotient remains at the speed of light.

Finally, the elevator stops at the center of Earth, and we examine ourselves. We are "weightless" because the surrounding mass of Earth pulls us equally in every possible direction. Earth's gravity is now being included in those pulls of gravity from all of those surrounding objects in the Universe that oppose each other and cancel out each other so that they do not contribute to the vector sum but create additional inertial force instead. This increased inertial force at the center of Earth does not pull at our bodies as a whole but pulls at each of the elementary particles of which we consist. Therefore, the orbits of the electrons bound to atomic nuclei are pulled farther away from those nuclei, which means that the Bohr radius is longer and the magnitude of the fine-structure constant, less.

As we ascend back to the surface, the reverse occurs, and, at the surface, the magnitudes of our bodily attributes are back where they were before we descended—well, not exactly. We appear not to have aged as much as those who remained behind at the surface. Of course these changes are minuscule because the mass of Earth is minuscule compared to the humongously more-massive objects in the Universe, which would affect us in humongously bigger ways.

B. Falling to Earth

We are high above Earth and appear to be motionless and weightless, apparently without any external forces being applied to our bodies. Yet, from every direction, a multitude of gravitational forces from all of the other objects in the Universe pull at the elementary particles, which make up our bodies. The vector sum of these forces accelerates us, along with all the other objects in the Solar System, at a particular rate in a particular direction. Also, we and Earth are attracting each other and accelerating toward each other. However, because Earth is considerably more massive than we, Earth appears to be immobile, and we, alone, are accelerating toward it.

Our rate of acceleration toward Earth is a function of two opposing forces—the gravitational force of attraction of Earth versus that of all the other objects in the Universe or, in other words—inertia. Earth pulls us in; inertia holds us back. Earth accelerates objects toward itself at the rate of a little less than 10 meters per second during each second. So, each of our 100-kilogram bodies has about 1,000 newtons of Earth's gravitational force pulling on it—$1,000 \ N \ = \ 100 \ kg \ \times \ 10 \ m \cdot s^{-2}$.

C. Falling in Sagittarius

We leave the Orion spiral arm of our Milky Way galaxy, which contains our Solar System, to enter the much larger and denser Sagittarius spiral arm, many light years away, where we visit a planet that possesses the same mass as does Earth. Suppose that all of the gravitational-force vectors that are applied to this earth-like planet possess magnitudes that are double those in the Solar System. This doubles the magnitude of the vector sum and the absolute acceleration, which we do not "feel" because we are free objects. Also, the magnitude of inertia doubles. Now, if the definition of the unit of inertial force remains the same, the magnitude of one newton doubles. Therefore, the acceleration of gravity at the surface of this earth-like planet would be about 5 meters per second during each second rather than 10— half that of our Earth. Other phenomena would be different too—we would move much slower, age much slower, think much slower, and be much bigger than on Earth. Again, we do not notice these differences because everything else around us also is different by the same amount.

D. Super-massive object

We are free objects, "weightless" somewhere in outer space, perhaps in the void between the Orion and Sagittarius spiral arms of our Milky Way galaxy. We appear to be motionless, but we know full well that we are in free-fall, accelerating in a particular direction. Yet, we cannot discern that direction, but we know that, as free objects, we are following a gravity line. We notice that stars and galaxies appear all around us as very faraway points of light, which, over a long period of time, appear not to move. This seems to mean that we are not rotating.

Suddenly, as if by magic and unbeknownst to us, an invisible super-massive object manifests itself nearby, abeam of our gravity line. One would believe that when the tremendous amount of speed-of-light gravitons from that humongous mass would hit us, we would be violently jerked 90° off our gravity line toward that super-massive object and rotated that 90°. However, we notice nothing amiss, nothing different. The points of light remain in their original positions—we have not rotated. We feel no extra forces—we are still "weightless." Yet, in reality, our gravity line has made a sharp 90° turn and points to the super-massive object. Also, our absolute acceleration toward that huge mass is, in itself, *huge* and dominates the accelerations caused by the gravitational pulls from the other objects, which are light-years away.

E. Alone in the Universe

We are somewhere in the Universe. It does not matter where because, by some magic, all objects in the Universe have vanished except for the two of us. We exist, but nothing else does. Of course, we are motionless (except for waving our arms and legs about), but we possess neither velocity nor acceleration; we are not in free-fall because no external forces act upon us—neither gravitational, inertial, nor mechanical. Then, we fire some rockets, which are attached to us. Supposedly, we are accelerated, but we can only detect acceleration when inertial forces exist, which, in our case, they do not. Also, acceleration creates velocity, but velocity only exists in relation to other objects, and, in our "empty" Universe, no other objects exist. So, even though our rockets are blasting away, we appear not to accelerate; we appear not to move; we hear the rockets roaring away, but nothing appears to happen—we feel no forces acting upon our bodies.

Yet, by being in an empty Universe, something has happened to our bodies: We have shrunk because no inertial forces are pulling the electrons away from the nuclei of the atoms in our bodies. Speculatively, the magnitude of Bohr's radius has shrunk to $r_B = \lambda_e (2 \pi \alpha)^{-1}$, where now $\alpha = 1$ instead of $1/137$. This means that we are smaller; we age faster; our thought processes are faster, yet we are aware of none of these changes because no relative comparisons are possible. Questions arise: If we are the only objects in the Universe, how big is the Universe now? When all objects except ourselves vanish, does the space that they had occupied vanish as well?

F. Orbiting Earth

We are now back in the normal Universe, returning to Earth from outer space. Our goal is to go into orbit about the Earth. One might think that to do so requires us to pass abeam of Earth a particular distance away from it and at a particular velocity relative to it. Apparently, this is not necessary to be very exact because, in nature, an innumerable number of orbiting stars, planets, moons, and comets do exist—so many so that to become a satellite seems not to be too difficult to achieve. However, if we want to enter a *near-circular* orbit at a *desired* distance away from Earth, our velocity and trajectory must be precise. If they are not, one of three results would oc-cur—slam into Earth, bypass Earth completely, or enter a stretched-out elliptical orbit much like that of a comet about the Sun. Surprisingly, one of a wide range of velocity-trajectory pairs would put us in orbit about Earth, albeit in one of various elliptical orbits, some quite eccentric. However, that is better than slamming into Earth or "flying" by it.

   In orbit, we are still in free-fall and "weightless." The gravitational pull of Earth becomes a centripetal force upon us because it continuously changes direction as we revolve about Earth. We do not spiral down to Earth because outward-directed inertial forces counteract Earth's gravity.

G. Landing on Earth

We are high above Earth in "weightless" free-fall, heading toward its center as indicated above in Section III.B. We notice that at an altitude of 1,000,000 meters above Earth, we approach it at a speed of 1,000 meters per second, and, if we do nothing, Earth's gravity would increase that speed by about another 10 meters per second during each succeeding second (ignor-ing wind resistance near Earth). We fire our retro rockets to *exactly* counter-act the pull of Earth's gravity, which applies about 1,000 newtons of force upon each of our 100-kilogram bodies. This external mechanical force mimics the force of gravity that would be applied to our bodies if we were standing on the surface of Earth because the acceleration would also pos-sess a magnitude of one g. That magnitude of force stops our acceleration, but we are still approaching Earth at a constant 1,000 meters per second (2,200 miles per hour). We throttle up the rockets to apply more force "up-ward" than Earth's gravity applies "downward," which, little-by-little, decreases our rate of descent until we touch the ground at zero speed rela-tive to it.

H. Being tiny

We have been used to having our bodies composed of a multitude of elementary particles—billions of them. Now, we find that we each consist of only one elementary particle—an electron. We are not real electrons because real ones cannot think or observe their environment, and we can—it is magic. We want to discover the smallest-possible magnitude of each of our dimensional attributes and, because we are elementary particles, this magnitude is a *quantum* one. This means that it cannot be separated into smaller magnitudes, and any bigger one must be *integral* (whole-number) multiples of the quantum magnitude—as the popular saying goes: Nature abhors a vacuum, but, on the quantum level, it also abhors fractions.

We are interested in discovering our *quantum* magnitude of each of five of our dimensions—mass, temperature, charge, length, and time. Two are easy to find; they are the magnitudes of the electron mass and charge. After some trial-and-error guessing, the other three appear to be the magnitudes of the electron Compton wavelength, the lifetime of the virtual electron, and the threshold temperature of the electron. Of course, the threshold temperature is not a quantum temperature but a *very hot* maximum temperature above which an electron cannot manifest itself. Our quantum length divided by our quantum time gives the speed of light.

These magnitudes are the *unidimensional quantum attributes of the electron*, which can be used in an electronic system of unit measures (see Section II.H. in *Part II: The Calculations*) provided that they are each assigned a value of unity (one). You might well ask: Why not use the historic *Système international* (SI) metric system of unit measures, which, apparently, has served us so well up to now? Frankly, it has failed us with regard to quantum measurements because it is based upon humanoid and terrestrial magnitudes rather than those of elementary particles. This has forced us to *create* (not discover) constants of proportionality, which we have erroneously and, perhaps, vainly baptized to be *fundamental universal physical constants of nature.* When using the *unidimensional quantum attributes of the electron* as units of measure upon formulating equations that pertain to phenomena of electrons, these "constants of proportionality" vanish or, rather, they acquire values of unity and, as mathematical *factors*, can be removed from the equations. These rationalized or "sanitized" equations indicate their meanings better than they do when they contain constants of proportionality. We will see more about this later in *Part Two: The Calculations*. For now, we observe how our quantum attributes affect our movements.

In our lowest level of excitation, we appear to constantly leap back and forth (vibrate) over a distance of one quantum length (nature forbids us to shorten this "leap"). One traversal takes one quantum time period; therefore, we are either immobile or moving at the speed of light. Yet, we appear to vanish during each light-speed traversal—called a *quantum* leap, which, ironically, the public erroneously considers to be a "big" leap. In *Part Two: The Calculations*, we will see how a quantum leap would also happen to us if we were bound to the nucleus of an atom, when we would leap from one "orbit" to another when we get excited (see Section III.I. in *Part II: The Calculations*) and how a quantum leap seems to apply to Werner Heisenberg's uncertainty principle (see Section III.B. in *Part II: The Calculations*), and to Thomas Young and George Airy's parallel lines and concentric rings, respectively (see Section IV.B.1. in *Part II: The Calculations*).

At higher levels of excitation, caused by a force from a particular direction, we appear to travel in jerks and pauses in that direction, the ratio between the occurrences of the jerks and pauses, depending upon the strength of the force. We travel one quantum length at a time, which takes one quantum time period to occur. As the magnitude of the force increases, more jerks occur and fewer pauses occur.

The absolute highest level of excitation occurs when we meet positrons. We each mate with a positron and transform ourselves into photons. Together, we still travel one quantum length at a time, which takes one quantum time period, but the pauses vanish, so we always travel at the speed of light, without any pauses—we are light, itself!

I. Summary

In our wanderings through the cosmos, we discovered that particular phenomena appear to be true. They are:

- Inertial mass and gravitational mass are one and the same phenomenon.
- Inertial force and gravitational force are one and the same phenomenon.
- Inertia is aggregate gravity.
- Classical gravity is discrete gravity.
- The magnitude of inertia varies throughout the Universe.
- The greater the magnitude of inertia, the larger matter is and the slower it ages.
- The greater the magnitude of inertia, the slower matter accelerates when an external force is applied to it.
- The magnitude of inertia in the vicinity of the Solar System is about 137 times greater than that in an otherwise empty universe.
- The Bohr radius of a hydrogen atom in the vicinity of the Solar System is about 137 times greater than that of one that is the only matter in an otherwise empty universe.
- The magnitude of inertial force at a point in space is proportional to the magnitude of the aggregate gravitational-force vectors applied there.

Part Two:

The Calculations

I. INTRODUCTION

Requisites for understanding the calculations in this essay are few. We need only to know a little about the basics of high-school algebra and physics, scientific (exponential) notation, and the metric system of unit measures. We will not even encounter calculus. Of course, we will acquire other fresh knowledge as we read along.

The calculations presented here were obtained by using dimensional analysis and algebraic manipulation of existing quantum-based proportions, equations, and formulas, which experiments by other persons in the past appeared to be correct. Originally, the results were obtained through either extrapolations, trial-and-error, or plausible guesses. Then, these results were correlated with established models and interpretations, based upon empirical experimental data, such as from George Airy's pinhole experiment, Thomas Young's double-slit experiment (see Section IV.B.), Werner Heisenberg's uncertainty principle (see Section III.B.), Max Planck's blackbody equation (see Section III.J.), Albert Einstein's photo-emission interpretation, Niels Bohr's model of the hydrogen atom (see Section III.I.), Jacob Bekenstein and Stephen Hawking's equation for the entropy of a Black Hole (see Section III.P.13.), et cetera.

The apparent correctness of these results prompted their extrapolation into subjective speculations as to the reality of particular aspects of the physical environment. These speculations were based upon those results that proved to be mathematically correct and because any other interpretation imaginable seemed to be less plausible, especially when assuming that, on the quantum level, reality should be simple, rational, and symmetrical.

In Niels Bohr's model of Ernest Rutherford's hydrogen atom (ignoring Arnold Sommerfeld and Paul Dirac's improvements to that model), *all* of the attributes of that atom's bound electron contain the fine-structure constant to the *exclusion of all other constants* except π [see Table XI under the column heading labeled **SE value (n = 1)**]. The magnitude of the ground-state orbit length of the electron is about 137 times longer than the quantum (wave)length λ_e of the electron. This seems to mean that the magnitude of the *fine-structure constant is inversely proportional to that of the linear size of matter.*

The methodology of quantum physics is simplified by factorizing each *fundamental universal physical constant of nature* into a combination of unidimensional quantum attributes of the electron. Some of these "rationalized" constants also contain π, an integer, or the fine-structure constant.[⊥] Among the affected constants are: Max Planck's h (see Eq. 3), Ludwig Boltzmann's k (see Eq. 97), Charles de Coulomb's permittivity ϵ_0 (see Eq. 13), André Ampère's permeability μ_0 (see Eq. 31), Isaac Newton's gravitational G (see Eq. 50), Armand Fizeau's speed-of-light c (see Eq. 84), and most any constant that occurs in an equation about quantum phenomena of the electron. See Tables VI and VII for more-complete lists of these constants.

The use of a system of unit measures, such as metric (mksA or cgs), that is not compatible with the quantum attributes of the electron creates the need for "constants of proportionality." For equations that pertain to the electron, I create a *quantum* system of unit measures (see Sections II.G.1. and II.H.). This system uses the magnitudes of the electron's unidimensional quantum attributes for its units of measure. When using this quantum system of units, most of the *fundamental universal physical constants of nature* that relate to the electron acquire a value of unity and, as arithmetic factors in an equation, in essence, vanish (see Table VI).

The fine-structure constant is one of the factors in the permittivity e_0, permeability μ_o, and gravitation G force constants (see Tables VI and VII). Obviously, this occurs because the Coulomb-, Ampère-, and Newton-force equations (see Eqs. 14, 29, and 49, respectively) were formulated using an inertial-force-based unit of force, the *newton*, (see III.D.1.) rather than an electromagnetic- or gravitational-force-based one.

Section II.G.1. presents the quantum attributes of the electron, which are its mass m_e,[2] threshold temperature k_e,[3] electronic charge q_e,[4] Compton wavelength λ_e,[5] and lifetime of a virtual electron t_e.[6] However, the electron's threshold temperature is not a quantum magnitude but a *maximum* one.

In Section II.H., using the same naming convention as that for *le Système international* (SI) of unit measures, I create *le Système électronique* (SE) for the electron. Combining the unidimensional attributes as factors in mathematical expressions creates a multitude of multi-dimensional quantum attributes of the electron. Tables I and II list these attributes including their SI values for comparison and verification. Tables VI and VII list a selection of historical fundamental universal physical constants of nature and the quantum attributes that are the factors of which each of these constants is composed.

II. DEFINITIONS

The following definitions will help us to understand the terminology of the proportions and equations that we will encounter later.

A. Letter-code case

A factor code that represents a *quantum* attribute of an elementary particle is represented by a lower-case letter. An integral (whole-number) multiple of a quantum attribute uses an upper-case letter. For example, the F_e code represents the magnitude of an integral multiple of f_e, which is the quantum of electromagnetic force.

B. Dimensionless value

A code, which normally represents a factor that possesses a number followed by units of measure but has those units transferred to another factor within the equation, is delimited by vertical bars. For example:
$h = 6.626 \times 10^{-34}$ kg·m^2·s^{-1} is Planck's constant with an SI value, including its SI units, which possess three dimensions—mass, length, and time; however, $|h|_m = 6.626 \times 10^{-34}$ is a dimensionless number, and the subscripted m indicates that it is an SI value. The dimensionless number, $|h|_e = 1$, and the subscripted e indicates that it is an SE value (see Section III.A.).

C. Mantissa accuracy

In this essay, to save space and unnecessary complication, I use only four digits to display the mantissas of the values in the text and tables; however, the calculations behind these displays use the most-accurate mantissas available. Most of them were obtained from the http://physics.nist.gov/cuu/Constants/index.html Internet site of the *National Institute of Standards and Technology (NIST)*. The *basic* values were stored in an HP-28C calculator to their highest level of accuracy available (range: plus and minus $1.00000000000 \times 10^{-499}$ to $9.99999999999 \times 10^{499}$) for calculating the *derived* values that are listed in this essay. If you wish to check my calculations, do *not* use the four-digit mantissas, which I display in this essay, but use the values that *NIST* specifies, otherwise the roundoff errors will be big.

D. Masstron

What I call a *masstron* is a hypothetical but plausible elementary particle of matter of my own invention, which fills a big "unknown" when calculating the phenomena of gravitational force. This big-unknown concept is similar in nature to Dmitri Mendeleev's "missing" but plausible elements in his periodic table of the elements or of how the unknown but plausible planets, Neptune and Pluto, were discovered by the orbital irregularities of the other known planets as detected by many astronomers from Urbain Le Verrier to Clyde Tombaugh. The masstron is to gravitational force what the electron is to electromagnetic force. The graviton would mediate the masstron much like the photon mediates the electron. Historically, calculations concerning gravitational force have left out the reality of the quantum attributes of an elementary particle of matter acted upon by gravitons; however, the invention of the Planck values seems to indicate that they were intended to be the attributes of an elementary particle that was hoped to have been found in the future. In essence, the magnitudes of the quantum attributes of the masstron have been discovered, but experimental proof of its existence seems to be elusive.

In addition to *le Système électronique* (SE) of unit measures for the electron, I also create *le Système gravitatif* (SG) for the masstron and *le Système protonique* (SP) for the proton (see II.H.).

E. Subscript conventions

To assure clarity of expression in an equation displayed in this essay, for a mathematical-factor code, I use a subscript, which can contain up to three characters. From left to right, these place-holder characters are arbitrarily labeled x, y, and z. The position of each character within the subscript specifies a particular property for its factor code, which can represent either an *attribute* or a *force*.

1. First subscript character (x)

For an *attribute* code, the first subscript character x specifies the elementary particle to which the attribute pertains, such as e for electron, p for proton, and g (gravity) for masstron. Examples are: λ_p is the quantum length of the proton (proton Compton wavelength); m_e is the mass of the electron; and m_g is the quantum mass of the masstron, which magnitude is *close* to that of the historical Planck mass (see Section III.I.8. and Table V for the explanation of the slight difference between these two magnitudes).

For a *force* code, the first subscript character x specifies the type of force, which can be either i for inertial, e for electronic (or, as the case requires, electromagnetic), m for magnetic, or g for gravitational. For example, F_g represents a whole-number multiple of the quantum of gravitational force f_g.

2. Second subscript character (y)

The second subscript character y specifies the system of unit measures used and can assume a value of either m for metric (SI), e for electromagnetic (SE), p for protonic (SP), or g for gravitational (SG). For example, m_{em} represents the magnitude of the mass of the electron expressed in metric (SI) values, and F_{gg} represents a magnitude of gravitational force expressed in SG values.

3. Third subscript character (z)

The third subscript character z specifies the system of unit measures used to display the units that are specified by the second subscript character. For example, F_{ggm} represents a magnitude of gravitational force, which is calculated using masstron attributes whose magnitudes are displayed in SI values. When the second and third subscript characters would specify the same system of unit measures, the third character is often not specified.

F. Proportions and equations

The mathematical expression on one side of either a proportion sign or an equal sign can mathematically represent an action (or a cause) that is applied to a particular physical system while that on the other side can represent the related reaction of (or effect upon) that physical system.

1. Proportion

To create a proportion, both the action and reaction in an experiment about physical processes must be repeatedly measured as the magnitudes of the action's factors are changed before each repetition or "run" of the experiment. When, in each case, the magnitudes of the action and reaction change by the same percentage (or factor), the proportion is valid. If this is not so, the proportion's factors must be reärranged or deleted, or other factors must be introduced until a valid proportion is created. The action and reaction of a proportion need possess neither the same dimensions, units of measure, values, nor magnitudes.

2. Constant of proportionality

The insertion of the appropriate constant of proportionality into a proportion converts it into an equation. This ensures that the action and reaction possess the same dimensions and magnitudes, which an equation requires.

For a generic example, let the action A be proportional to the reaction B, where A and B possess different values, dimensions, magnitudes, and units of measure; therefore, $(A \propto B)$. When the constant of proportionality is inserted into the numerator to the right of the proportion sign, that constant's value is $(A \cdot B^{-1})$, which changes the proportion $(A \propto B)$ into the equation $[A = (A \cdot B^{-1}) \ B]$ and, by cross cancelling $(B^{-1} \cdot B)$, gives the true equation $(A = A)$. Suppose that the action A changes by $+n\%$. The reaction B would also change by the same $+n\%$ to give the proportion $[A(1+n\%) \propto B(1+n\%)]$. The corresponding equation would become $\{[A(1+n\%)] = [A(1+n\%)] \ [B(1+n\%)]^{-1} \ [B(1+n\%)]\}$, which, after cross cancelling gives $[A(1+n\%)] = [A(1+n\%)]$. If the two percentages had been different, A would not have been proportional to B, and the equation could not have been formed.

This procedure is how the permittivity- and gravitational-force constants were created. See Section III.F.2. for the explanation about creating (defining) the permeability constant.

3. Equation

An equation that contains a constant of proportionality had been formed from a proportion that represented the mathematical connection between a physical action and its resultant reaction. The action and reaction represented in an equation must possess the same dimensions, and magnitudes. The constant of proportionality ensures this. However, an equation need contain neither the same values nor units of measure on either side of the equal sign. For example, the $(1 \ m = 100 \text{cm})$ equation possesses the same dimensions and magnitudes but neither the same values nor units of measure. However, the insertion of a dimensionless unit-conversion factor of unity magnitude *can* force the equation to possess equal values and units of measure. The above $(1 \ m = 100 \ cm)$ equation can become the trivial $(1 \ m = 1 \ m)$ equation by multiplying $(100 \ cm)$ by the $[1 \ m \ (100 \ cm)^{-1}]$ unit-conversion factor, and cross cancelling $[100 \ cm \ (100 \ cm)^{-1}]$.

The advantage of an equation over a proportion is that, to determine the value of the reaction, it need not be measured. After the value of the constant of proportionality is established through experimentation, only insertion of the values of the action's factors into the equation need be necessary to calculate the reaction's values.

G. Quantum attributes of elementary particles

This section presents the SI values of the unidimensional quantum attributes of the electron, the proton, and the masstron. Tables I through IV show, more completely, these attributes and the relationships between the attribute magnitudes of these three elementary particles.

1. Quantum attributes of the electron

According to *NIST* at http://physics.nist.gov/cuu/Constants/index.html, the unidimensional quantum attributes of the electron, their codes, and SI values to their current best-known accuracies, except for the threshold temperature and lifetime of a virtual electron, which I cannot find in the *NIST* listing, are:

Electron quantum attribute	Code	SI value
mass	m_e	$9.1093826 \times 10^{-31}$ kg
threshold temperature	k_e	5.9298894×10^{9} K
electric charge	q_e	$-1.60217653 \times 10^{-19}$ C
Compton wavelength	λ_e	$2.426310238 \times 10^{-12}$ m
lifetime of a virtual electron	t_e	$8.093299792 \times 10^{-21}$ s

I obtained the value of the electron threshold temperature by dividing $(m_e\, c^{\,2})$ by Boltzmann's constant and that of the lifetime by dividing the electron Compton wavelength by the speed of light. The values of both of these divisors are listed by *NIST*.

2. Quantum attributes of the proton

According to *NIST*, the unidimensional quantum attributes of the proton, their codes, and SI values to their current best-known accuracies, except for the threshold temperature and lifetime of a virtual proton, which I cannot find in the *NIST* listing, are:

Proton quantum attribute	Code	SI value
mass	m_p	$1.67262171 \times 10^{-27}$ kg
threshold temperature	k_p	$1.08881522 \times 10^{13}$ K
electric charge	q_p	$+ 1.60217653 \times 10^{-19}$ C
Compton wavelength	λ_p	$1.3214098555 \times 10^{-15}$ m
lifetime of a virtual proton	t_p	$4.4077488283 \times 10^{-24}$ s

I obtained the value of the proton threshold temperature by dividing ($m_p c^2$) by Boltzmann's constant and that of the lifetime by dividing the proton Compton wavelength by the speed of light. *NIST* lists the values of both of these divisors.

Although I found the SI values of these attributes of the electron and proton at *NIST* and in literature at physics libraries, I did not find there that they are recognized as their *unidimensional quantum* attributes.

3. Beta ratio

Historically, the *Beta* ratio has been recognized as only the ratio between the magnitude of the *mass* of the proton over that of the electron. Nowhere in the physics literature did I find it to be considered as the ratio between the magnitudes of all the other corresponding quantum attributes of the electron and proton. When I discovered that this was so, I became excited because it meant that the quantum attributes of the electron are linked to those of the proton by *only one* mathematical factor—the "ubiquitous" beta ratio, which possesses a magnitude of $\beta = 1836.15267261....$

Notice in the above two tables that the ratio between the magnitudes of each of the electron and proton's corresponding unidimensional quantum attributes, except for electric charge, is the *same*. However, the magnitudes of the proton mass and threshold temperature are $1836.15267261...$ times *greater* than those of the electron while those of the proton Compton wavelength and virtual-proton lifetime are $1836.15267261...$ times *less* than those of the electron, which maintains the speed-of-light quotient between the wavelength and the lifetime. Apparently, this means that as the mass of an elementary particle *increases* by a particular factor, the distance of its "quantum leap" *decreases* by that same factor. Electron-to-proton comparison is like that of cheetah-to-elephant seal—the light cheetah covers a long distance at each leap while the heavy elephant seal covers a few inches at each heave, yet their momenta are nearly equal.

This lead me to speculate that possibly other elementary pairs might possess the same phenomenon, and, after much juggling of mathematical figures, I found this to be true. This was mind boggling because it lead to still another important discovery—the simple, yet elegant, linkage between electromagnetic- and *gravitational* force. I call this linkage the *Gamma* ratio, which mirrors the logic of the *Beta* ratio. It is the ratio between the magnitudes of each of the electron and masstron's corresponding unidimensional quantum attributes.

4. Gamma ratio

By extrapolating the concept of the *Beta* ratio into gravitational phenomena and after much trial-and-error, I discovered the *Gamma* ratio, which is $\gamma = 2.041 \times 10^{21}$. Yes, you read that right—ten to the *twenty-first* power. The magnitudes of the masstron mass and threshold temperature are 2.041×10^{21} times *greater* than those of the electron while those of the masstron Compton wavelength and virtual-masstron lifetime are 2.041×10^{21} times *less* than those of the electron. See Section II.G.5. (below). This makes the cheetah-elephant seal comparison look mild—a hummingbird-to-whale comparison will not, in this case, even come close. Yet, the "massless" graviton, which would mediate the masstron, appears to consist of a masstron paired with its conjugate antimasstron at the speed of light. Even though the masstron appears to possess a relatively huge mass, and the antimasstron, a huge antimass, they combine into a zero-mass graviton.

By squaring the *Gamma* ratio, I was relieved to see something familiar, $\gamma^2 = 4.166 \times 10^{42}$, which is the magnitude of the historical electro-magnetic to a gravitational-force ratio. Then, I discovered that the magnitudes of the unidimensional quantum attributes of the masstron are *very* close to those of the historical Planck values, which fact reënforces the concept that the masstron *really* does exist. Table V shows the slight differences between the magnitudes of the masstron's quantum attributes and their related historical Planck values. These differences are the $\alpha^{\frac{1}{2}}$ and (2π) factors. Section III.G.8. explains that these differences exist because of the historical formulation of Newton's gravitational-force equation and its included G constant, which is a factor that was used in calculating the Planck values.

5. Quantum attributes of the masstron

You might be skeptical about the existence of an elementary particle of matter—the *masstron*—because, perhaps, I am the one who *first believed that it does exist*. Yet, while studying the quantum aspects of gravitational force, an elementary particle like the masstron begs to be recognized. De Coulomb's electron-force equation has the electron, but Newton's gravitational-force equation, up until now, appears to have *nothing*. Symmetry demands that the masstron exist and that its unidimensional quantum attributes possess the values listed in the table below. These attributes of the masstron, their codes, and SI mantissa values truncated to four significant digits are:

Masstron quantum attribute	Code	SI value
mass	m_g	1.859×10^{-9} kg
threshold temperature	k_g	1.210×10^{31} K
Compton wavelength	λ_g	1.188×10^{-33} m
lifetime of a virtual masstron	t_g	3.965×10^{-42} s

Notice that the values of the masstron mass and threshold temperature are *larger* than those of the electron by a factor of $\gamma = 2.041 \times 10^{21}$, while the values of the masstron Compton wavelength and lifetime of a virtual masstron are *less* than those of the electron by that *same* factor of $\gamma = 2.041 \times 10^{21}$.

6. Discussion

The quantum attributes of an elementary particle appear to be the smallest possible magnitudes for these attributes. When an elementary particle moves, it seems to vanish during its quantum-time attribute to reäppear its quantum-length attribute away. For example, the electron vanishes, then, after a time of t_e, it reäppears a λ_e distance away. This means that elementary particles appear to always make speed-of-light "quantum leaps." This is more evident in atoms when bound electrons make quantum leaps from one orbit to another or when streams of photons or electrons refract at whole-number multiples of a quantum angle in Augustin Fresnel, Thomas Young, and George Airy's experiments (see Section IV.B.2.).

H. Système électronique, protonique, et gravitatif

Using the same naming convention as that for the historic *Système international* (SI) of unit measures, I create *le Système électronique* (SE) of unit measures for the electron (see Table I), *le Système protonique* (SP) for the proton (see Table III), and *le Système gravitatif* (SG) for the masstron (see Table IV). This is done by assigning a value of unity to each of the unidimensional quantum attributes for these elementary particles. For example, the SE system of unit measures for the electron uses $m_e = 1$, $k_e = 1$, $q_e = -1$, $\lambda_e = 1$, and $t_e = 1$ as shown in Table I.

III. MATHEMATICAL CALCULATIONS

Each subsection that follows shows the mathematical calculations for converting a particular fundamental universal physical constant of nature into a collection of factors, each of which is either π, an integer, the fine-structure constant, or a unidimensional quantum attribute of the elementary particle to which the constant pertains.

Fear not, these calculations are not really algebraic but arithmetic, except that codes replace real values when appropriate. No mathematical *terms* exist in this essay—neither addends nor subtrahends—only mathematical *factors* exist, which, by definition, are either multipliers or divisors, or powers of them. Of course, a divisor is indicated by a negative power, which eliminates the need for built-up fractions. This simplifies the presentation of the mathematical material.

A. Planck's constant

In working with cavity radiancy in 1900, Max Planck introduced the world to what became to be known as Planck's constant by isolating it in both of Wilhelm Wien's radiation constants, c_1 and c_2, which were needed to create an equation that mirrored the results acquired through experiments. In 1905, Albert Einstein used it in his work with the photoelectric effect. Thereafter, to convert any *quantum* proportion, which had a dimensional difference of $M\ L^2\ T^{-1}$ between its left and right components, into an equation, the insertion of Planck's constant as the constant of proportionality became necessary. For information about Max Planck's work with cavity radiancy, see Section III.J. The SI value of Planck's constant[7] is:

$$h_m\ =\ 6.626\ \times\ 10^{-34}\ \text{kg·m}^2\text{·s}^{-1}, \tag{1}$$

which value is listed in Tables VI and VII. Planck's constant can be factorized into a combination of more-basic constants, which are quantum attributes of the electron. These attributes are arranged in accordance with the dimensions of Planck's constant: $M\ L^2\ T^{-1}$. Therefore (see Section II.G.1.),

$$\begin{aligned}
h_m\ &=\ (9.109\ \times\ 10^{-31}\ \text{kg})\\
&\times\ (2.426\ \times\ 10^{-12}\ \text{m})^2\\
&\times\ (8.093\ \times\ 10^{-21}\ \text{s})^{-1}\\
&=\ 6.626\ \times\ 10^{-34}\ \text{kg·m}^2\text{·s}^{-1}.
\end{aligned} \tag{2}$$

1. Using SE units

Using the quantum attributes of the electron as the SE units of measure and expressing Planck's constant using these SE units gives:

$$h_e \; = \; (1 \; m_e) \; (1 \; \lambda_e)^2 \; (1 \; t_e)^{-1} \; = \; 1 \; m_e \cdot \lambda_e^2 \cdot t_e^{-1} \; = \; 1 \; E_e \cdot t_e \; . \quad (3)$$

As seen above, the magnitude of Planck's constant can be expressed as the product of the magnitude of the rest-mass energy of the electron and the magnitude of the lifetime of a virtual electron.

2. Discussion

Planck's constant is composed of factors, all of which can be the quantum attributes of either the electron, proton, or masstron. This is so because the changes in both the numerator and denominator factors cancel out in each case. In any one of these cases, its quantum value is unity; therefore, in reality, Planck's constant *does not exist as a fundamental constant.*

B. Heisenberg's uncertainty principle

Now that we know the *real* definition of Planck's constant, we can more-precisely explain Werner Heisenberg's uncertainty principle. In 1927, he speculated that *precise* simultaneous measurements of *all* of the attributes of an elementary particle on the quantum scale are impossible because of their discontinuousness between whole numbers. Even the most precise measuring instrument can measure only down to magnitudes of *unity* and no further. On the quantum level, only whole-numbered attributes exist- fractional attributes are forbidden by nature.

1. Using SI units

Using SI units, Werner Heisenberg's uncertainty principle [8] was formulated as $[\Delta x \; \Delta p_x \; \geq \; h]$, which can be read as: The mathematical product of an elementary particle's changes in location and momentum in the x direction cannot be less than the magnitude of Planck's constant.

2. Using SE units

When using SE units, the uncertainty principle can be stated more precisely as $(\Delta x \; \Delta p_x \; \geq \; \lambda_e \, p_e)$, where $p_e \; = \; m_e \cdot \lambda_e \cdot t_e^{-1}$, the quantum-electron momentum. The principle can be read as: The mathematical product of an electron's changes in location and momentum in the x direction cannot be less than the product of its quantum-length attribute and its quantum-momentum attribute.

Also, $(\Delta E \ \Delta t \geq h)$, in SE units, can be expressed as $(\Delta E \ \Delta t \geq E_e \cdot t_e)$, where $E_e = m_e \cdot \lambda_e^2 \cdot t_e^{-2}$, the rest-mass energy of the electron.

Similarly, $(\Delta f \ \Delta t \geq 1)$, in SE units, can be expressed as $(\Delta f \ \Delta t \geq t_e^{-1} \cdot t_e)$, where $t_e^{-1} \cdot t_e = 1$.

3. Using SG units

When using SG units, the uncertainty principle can be stated more precisely as $(\Delta x \ \Delta p_x \geq \lambda_g \cdot p_g)$, where $p_g = m_g \cdot \lambda_g \cdot t_g^{-1}$, the quantum-masstron momentum. The principle can be read as: The mathematical product of a masstron's changes in location and momentum in the x direction cannot be less than the product of its quantum-length attribute and its quantum-momentum attribute.

Also, $(\Delta E \ \Delta t \geq h)$, in SG units, can be expressed as $(\Delta E \ \Delta t \geq E_g \cdot t_g)$, where $E_g = m_g \cdot \lambda_g^2 \cdot t_g^{-2}$, the rest-mass energy of the masstron.

Similarly, $(\Delta f \ \Delta t \geq 1)$, in SG units, can be expressed as $(\Delta f \ \Delta t \geq t_g^{-1} \cdot t_g)$, where $t_g^{-1} \cdot t_g = 1$.

4. Discussion

Evidently, Heisenberg's uncertainty principle can be stated even more precisely than was possible in the past: Not only can the product of two attributes of an elementary particle not be less than the product of the magnitudes of their respective quantum attributes—the magnitude of a single attribute cannot be less than its quantum magnitude. Examples are: $\Delta x_g \geq \lambda_g$, $\Delta p_{xe} \geq p_e$, $\Delta E_g \geq E_g$, $\Delta t_e \geq t_e$, and $\Delta f_g \geq t_g^{-1}$. The interpretation and meaning of Werner Heisenberg's uncertainty principle are simplified by separating Planck's constant into its constituent quantum factors.

At a particular moment in time, the magnitudes of an elementary particle's attributes are fixed in every dimension. They only can change by a quantum attribute (make a *quantum leap*) when the next quantum moment in time arrives. Within a quantum time interval, nothing can occur because it is indivisible. In reality, quantum momentum does not exist because, when an elementary particle moves, it seems to vanish from the Universe to reäppear a quantum distance away after a delay of a quantum time period.

Dividing the magnitude of the quantum distance by that of the quantum time period, *for any elementary particle*, gives the speed of light.

Although all of the magnitudes of an elementary particle's quantum attributes are fixed at a particular moment in time, an attempt to measure them changes them because the magnitudes of the measurement signals cannot be less than those of the quantum attributes themselves.

C. Extranuclear forces

Various forms of phenomena create various types of extranuclear force. Pushing a mass creates inertial F_i force. Groups of electrons create electronic F_e force between themselves. Currents of electrons create magnetic F_m force between themselves. Groups of masses create gravitational F_g force between themselves. Historically, all of these forces were measured and calculated using the SI unit of force, the *newton* N, which is *not based upon any of those forces* but upon inertial force. This practice necessitates the insertion of the fine-structure constant into electromagnetic- and gravitational-force proportions to convert them into equations (compare Eqs. 17, 34, and 53 with each other), yet, historically, this was not recognized because nobody knew that the fine-structure constant was a factor in the force constants—now we know.

By using the quantum units of measure, the historical force constants vanish from force equations except for their fine-structure-constant factor. These force constants are:

- Charles de Coulomb's permittivity constant ϵ_0 for electronic force,
- André Ampère's permeability constant μ_0 for magnetic force, and
- Isaac Newton's gravitation constant G for gravitational force.

The predecessor to the SI (mksa) system of metric unit measures is the cgs system, which has two branches. The one used for electrical phenomena is the electrostatic-unit (esu) system, which, by definition, assigns a value of one to the *permittivity* constant. The one used for magnetic phenomena is the electromagnetic-unit (emu) system, which, by definition, assigns a value of one to the *permeability* constant. In this essay, I use *le Système international* (SI) of metric unit measures.

D. Newton's inertial force

More than three centuries ago, Isaac Newton[9] discovered that the magnitude of a gravitational or mechanical force that is applied to a free object is proportional to the product of the magnitude of that object's mass and acceleration, which acceleration occurs only while that force is applied to that object. This proportion is formulated as ($F_i \propto m$ a). At the time, the value of the force probably was *not* defined to be *equal* to the value of the product of the mass and acceleration, so in that case, Newton would not have created the current ($F_i = m$ a) equation, which is not really a complete equation because a constant of proportionality is lacking. Possibly, he created the ($F_i = k\ m$ a) equation instead, where k was the inertial constant, a constant of proportionality, which ensured that both sides of the equation would contain the same dimensions.

1. Inertial constant

Years later, the SI value of the inertial constant was *defined* to be unity; therefore, it is, unfortunately, not actually inserted into inertial-force equations. Yet, understanding why it does exist is important to the context of this essay.

Actually, F_i is an action or cause, and (m a) is a reaction or effect. The one is proportional to the other but *not* equal to the other. A force does not possess dimensions of mass times acceleration (M L T^{-2}). It possesses a dimension of all its own—a dimension of force F. This force dimension can be proportional, not only to the M L T^{-2} dimensions for inertial force, but to the Q^2 L^{-2} dimensions for electronic force, Q^2 T^{-2} for magnetic force, and M^2 L^{-2} for gravitational force, as we will see later.

After the invention of the SI metric system of unit measures, long after Newton's death, the SI unit of (inertial) force f_{im} was named the *newton* N and its unity value was defined to increase (or decrease) the velocity of a one-kilogram mass by one meter per second when applied to that mass for one second. This gives the equation for the definition of
[$f_{im} = 1\ N = k\ (1\ kg{\cdot}m{\cdot}s^{-2})$] as the SI unit of force, where k is the inertial constant and is equal to (1 N${\cdot}kg^{-1}{\cdot}m^{-1}{\cdot}s^2$), which gives a true equation of (1 N = 1 N). Because the value of k was unity, it was never used, and the current inertial-force equation became ($F_i = m$ a), although it is not a true equation where the dimensions on either side of the equal sign equal each other.

2. SE unit of inertial force

The SE unit of inertial force, $\boldsymbol{f}_{ie}$, is proportional to $m_e \cdot \lambda_e \cdot t_e^{-2}$. Its SI value is (see Section II.G.1. for the factor values):

$$\boldsymbol{f}_{iem} = (1 \ N \cdot kg^{-1} \cdot m^{-1} \cdot s^2) \ (9.109 \times 10^{-31} \ kg)$$
$$\times \ (2.426 \times 10^{-12} \ m) \ (8.093 \times 10^{-21} \ s)^{-2}$$
$$= 3.374 \times 10^{-2} \ N, \tag{4}$$

which value is used for comparative purposes later in this essay. See Table VIII.

3. SG unit of inertial force

The SG unit of inertial force, $\boldsymbol{f}_{ig}$, is proportional to $m_g \cdot \lambda_g \cdot t_g^{-2}$. Its SI value is (see Section II.G.5. for the factor values):

$$\boldsymbol{f}_{igm} = (1 \ N \cdot kg^{-1} \cdot m^{-1} \cdot s^2) \ (1.859 \times 10^{-9} \ kg)$$
$$\times \ (1.188 \times 10^{-33} \ m) \ \times \ (3.965 \times 10^{-42} \ s)^{-2}$$
$$= 1.405 \times 10^{41} \ N, \tag{5}$$

which value is used for comparative purposes later in this essay. See Table IX.

4. Inertial-force equation in SE units

The SE inertial-force equation is:

$$\boldsymbol{F}_{ie} = |M_e| \ |\boldsymbol{A}_e| \ \boldsymbol{f}_{ie} , \tag{6}$$

where:
    $|M_e|$ = a dimensionless whole number of $|m_e|$ magnitudes,
    $|\boldsymbol{A}_e|$ = a dimensionless whole number of $|\lambda_e \cdot t_e^{-2}|$ magnitudes, and
    $\boldsymbol{f}_{ie}$ = 1 SE quantum unit of inertial force.

5. Inertial-force equation in SG units

The SG inertial-force equation is:

$$F_{ig} = |M_g| \ |A_g| \ f_{ig} , \tag{7}$$

where:
  $|M_g|$ = a dimensionless whole number of $|m_g|$ magnitudes,
  $|A_g|$ = a dimensionless whole number of $|\lambda_g \cdot t_g^{-2}|$ magnitudes, and
  f_{ig} = 1 SG quantum unit of inertial force.

6. Ratio of the SG unit to the SE unit of inertial force

The factor γ^2 is the gravitational-to-electromagnetic quantum inertial-force ratio:

$$\gamma^2 = f_{igm} \ f_{iem}^{-1} = (1.405 \times 10^{41} \ N)$$
$$(3.374 \times 10^{-2} \ N)^{-1} = 4.166 \times 10^{42}. \tag{8}$$

See Eqs. 4 and 5 and Table X. γ^2 is also the historical electromagnetic-to-gravitational[10] force ratio.

E. De Coulomb's electronic force

In about 1785, Charles de Coulomb[11] discovered that two point charges (q_1 and q_2) of electrons a distance r from each other create an electronic force of repulsion F_e between themselves.

1. Electronic-force proportion

De Coulomb's electronic-force proportion possesses an action of dimensions: charge squared per length squared $Q^2 \ L^{-2}$ and a reaction of the dimension: electronic force F_e such that $F_e \propto q_1 \ q_2 \ r^{-2}$. Later, the proportion was revised to be $F_e \propto q_1 \ q_2 \ (4 \ \pi \ r^2)^{-1}$, where $(4 \ \pi \ r^2)$ is the area of the surface of an imaginary sphere, which is concentric to one of the point charges. The other point charge resides at a point upon that sphere's surface.

2. Permittivity constant in SI units

The permittivity constant ϵ_0 converts the electronic-force *proportion* into the electronic-force *equation*. An example *hypothetical* experiment shows how the SI value of the permittivity constant was determined (de Coulomb determined a different value because, at the time, the metric system of unit measures did not yet exist): Let us arbitrarily set up an action as a point charge of $q_1 = 2.3 \times 10^{-6}$ C separated from a second arbitrary point charge of $q_2 = 4.7 \times 10^{-6}$ C by a distance of $r = 25$ cm. In this experiment, actually measuring the electronic force of repulsion between the two charges would indicate a reaction of $F = 1.554$ N. The proportion is:

$$1.554 \text{ N} \propto (2.3 \times 10^{-6} \text{ C}) \ (4.7 \times 10^{-6} \text{ C})$$
$$\times \ [4 \ \pi \ (0.25 \text{ m})^2]^{-1} \tag{9}$$
$$\text{or}$$
$$1.554 \text{ N} \propto 1.376 \times 10^{-11} \text{ C}^2 \cdot \text{m}^{-2}. \tag{10}$$

If we were to change the magnitudes of the action's three arguments for each "run" of the experiment, the action's value would change by a particular percentage, and the reaction's value would change by that same percentage; therefore, no matter what the values of the action's three factors would be, dividing the value of the reaction by that of its corresponding action would yield the same value—a constant of proportionality—the reciprocal of the permittivity constant:

$$\epsilon_0^{-1} = (1.554 \text{ N}) \ (1.376 \times 10^{-11} \text{ C}^2 \cdot \text{m}^{-2})^{-1}$$
$$= 1.129 \times 10^{11} \text{ N} \cdot \text{C}^{-2} \cdot \text{m}^2. \tag{11}$$

The reciprocal is yielded because, historically, the permittivity constant was placed in the denominator of the action (right) side of the equation. The SI value of the permittivity constant [12] (*NIST* calls this constant the *electric constant*) is (see Table VI):

$$\epsilon_0 = 8.854 \times 10^{-12} \text{ N}^{-1} \cdot \text{C}^2 \cdot \text{m}^{-2}. \tag{12}$$

3. Permittivity constant in SE units

To obtain the permittivity constant using SE units, we can substitute the SE units for the SI units, which gives:

$$e_0 = (8.854 \times 10^{-12}) \ [(3.374 \times 10^{-2})^{-1} \ f_{ie}]^{-1}$$
$$\times \ [(1.602 \times 10^{-19})^{-1} \ q_e]^2 \ [(2.426 \times 10^{-12})^{-1} \ \lambda_e]^{-2}$$
$$= 68.518 \ f_{ie}^{-1} \cdot q_e^2 \cdot \lambda_e^{-2} = (2 \ \alpha)^{-1} \ f_{ie}^{-1} \cdot q_e^2 \cdot \lambda_e^{-2}. \qquad (13)$$

The 68.518 value is the reciprocal of twice the value of the fine-structure constant. So, now, we realize that the permittivity constant is not really a fundamental constant at all.

4. Electronic-force equation in SI units

We already know that the SI electronic-force equation is:

$$F_{em} = q_1 \ q_2 \ [e_0 \ (4 \ \pi \ r^2)]^{-1}, \qquad (14)$$

where:
  q_1 and q_2 each represents an electric charge in coulombs,
  r is the distance between their centers in meters, and
  e_0 is the permittivity constant in farads per meter or $(N^{-1} \cdot C^2 \cdot m^{-2})$.

5. Electronic-force equation in SE units

Separating the values from the units in the SI electronic-force equation (see Eq. 14), in preparation for creating the electronic-force equation in SE units, gives:

$$F_{em} = (|q_1|_m \ C) \ (|q_2|_m \ C)$$
$$\times \ [(8.854 \times 10^{-12} \ N^{-1} \cdot C^2 \cdot m^{-2}) \ (4 \ \pi \ |r^2|_m \ m^2)]^{-1}. \qquad (15)$$

Replacing the SI units with their SE units of the same magnitude gives:

$$F_{ee} = (|q_1|_e \ q_e) \ (|q_2|_e \ q_e)$$
$$\times \ \{[(2 \ \alpha)^{-1} \ f_{ie}^{-1} \cdot q_e^2 \cdot \lambda_e^{-2}] \ (4 \ \pi \ |r^2|_e \ \lambda_e^2)\}^{-1}. \qquad (16)$$

Note that, in magnitude but not in value, $(|q_1|_m \ C = |q_1|_e \ q_e)$ and $(|r^2|_m \ m^2 = |r^2|_e \ \lambda_e^2)$.

Cross cancelling units and simplifying gives:

$$\boldsymbol{F}_{ee} = 2 \; |q_1|_e \; |q_2|_e \; (4 \; \pi \; |r^2|_e)^{-1} \; (\alpha \; \boldsymbol{f}_{ie}), \qquad (17)$$

where entries delimited by vertical bars are dimensionless whole numbers, which possess SE values. The factors $(\alpha \; \boldsymbol{f}_{ie})$ are equal to the factor $\boldsymbol{f}_{ee}$ (see Table VIII), which is the SE unit of electronic force, because, historically, the value of the permittivity constant reflected the practice of using an inertial-force-based unit of force (the newton N) in defining its value rather than an electronic-force-based one. This means that the correct SE electronic-force equation is (see Eq. 22 and the section that follows this one):

$$\boldsymbol{F}_{ee} = 2 \; |q_1|_e \; |q_2|_e \; (4 \; \pi \; |r^2|_e)^{-1} \; \boldsymbol{f}_{ee}. \qquad (18)$$

For an explanation of why the number 2 appears at the beginning of the definition for $\boldsymbol{F}_{ee}$, see Section II.J.2. in *Part I: The Voyage*, which relates to gravitational force, not electromagnetic, but the principle is the same because they both possess spherical symmetry.

6. SE unit of electronic force

The SE unit of electronic force $\boldsymbol{f}_{ee}$ is the reaction to one electron q_e separated from another electron q_e by the SE unit of length λ_e such that $\boldsymbol{f}_{ee}$ is proportional to, but not equal to, $q_e^2 \cdot \lambda_e^{-2}$. The action is $q_e^2 \cdot \lambda_e^{-2}$, the reaction, $\boldsymbol{f}_{ee}$. All other possible magnitudes are positive, whole-number multiples of this electronic-force quantum. Its proportion is:

$$\boldsymbol{f}_{ee} \; \propto \; q_e^2 \cdot \lambda_e^{-2}. \qquad (19)$$

Using de Coulomb's historical SI electronic-force equation, which includes the permittivity constant as a factor (see Section III.E.4.), calculating the magnitude of $\boldsymbol{f}_{ee}$ gives a *preliminary* SI value of:

$$\begin{aligned}
\boldsymbol{f}_{eem} &= q_e^2 \; [4 \; \pi \; \epsilon_0 \; \lambda_e^2]^{-1} = (1.602 \times 10^{-19} \text{ C})^2 \\
&\times \; [4 \; \pi \; (8.854 \times 10^{-12} \text{ N}^{-1} \cdot \text{C}^2 \cdot \text{m}^{-2}) \; (2.426 \times 10^{-12} \text{ m})^2]^{-1} \\
&= 3.918 \times 10^{-5} \text{ N}. \qquad (20)
\end{aligned}$$

At first, the quantum electronic-force equation appears to be:

$$F_{ee} = |q_1|_e \ |q_2|_e \ |r|_e^{-2} \ f_{ee},\qquad(21)$$

where:

$|q_1|_e$ and $|q_2|_e$ each represent an integral number of electrons, and $|r|_e$ is the integral number of SE units of length between them.

However, spherical symmetry *requires* that the equation be:

$$F_{ee} = 2 \ |q_1|_e \ |q_2|_e \ (4 \ \pi \ |r|_e^2)^{-1} \ f_{ee},\qquad(22)$$

which places 2 and $(4 \ \pi \ r^2)$, the area of the sphere's surface (of points of equal force), into the equation.

Therefore, the *correct* magnitude of f_{ee} in the SI unit of force is (see Eq. 20 and Table VIII):

$$f_{eem} = \tfrac{1}{2} \ (4 \ \pi) \ (3.918 \times 10^{-5} \ N)$$
$$= 2.462 \times 10^{-4} \ N.\qquad(23)$$

Comparing Eq. 4 with Eq. 23 shows that the magnitude of the quantum of electron-based inertial force is $137....$ times greater than that of electronic force (see Table VIII):

$$f_{iem} \ f_{eem}^{-1} = (3.374 \times 10^{-2} \ N) \ (2.462 \times 10^{-4} \ N)^{-1}$$
$$= 137.....\qquad(24)$$

This electronic phenomenon gives one definition of the fine-structure constant. The magnetic- and gravitational-force phenomena give two more (see Eqs. 41 and 60).

7. Comparing quantum and classical electronic-force equations

In an example quantum electronic-force equation, arbitrarily let:

$|q_1|_e = 3.5 \times 10^{12}$ (dimensionless whole number of electrons),
$|q_2|_e = 2.0 \times 10^{10}$ (dimensionless whole number of electrons), and
$|r|_e = 1.8 \times 10^{6}$ (dimensionless whole number of quantum electronic units of length).

These magnitudes, using SI values, are:

q_1 = (3.5 × 10^{12}) (1.602 × 10^{-19} C) = 5.607 × 10^{-7} C,
q_2 = (2.0 × 10^{10}) (1.602 × 10^{-19} C) = 3.204 × 10^{-9} C,
and
r = (1.8 × 10^6) (2.426 × 10^{-12} m) = 4.367 × 10^{-6} m.

The quantum equation in SE units is:

$$\boldsymbol{F}_{ee} = 2 \ (3.5 × 10^{12}) \ (2.0 × 10^{10})$$
$$× \ [4 \ \pi \ (1.8 × 10^6)^{\ 2}]^{-1} \ \boldsymbol{f}_{ee} = 3.438 × 10^9 \ \boldsymbol{f}_{ee}. \quad (25)$$

For later comparison, replacing $\boldsymbol{f}_{ee}$ with its SI value (see Eq. 23) gives:

$$\boldsymbol{F}_{eem} = (3.438 × 10^9) \ (2.462 × 10^{-4} \ N)$$
$$= 8.462 × 10^5 \ N. \quad (26)$$

The classical equation in SI units is:

$$\boldsymbol{F}_{em} = [(5.607 × 10^{-7} \ C) \ (3.204 × 10^{-9} \ C)]$$
$$× \ [4 \ \pi \ (8.854 × 10^{-12} \ N^{-1}{\cdot}C^2{\cdot}m^{-2}) \ (4.367 × 10^{-6} \ m)^2]^{-1}$$
$$= 8.462 × 10^5 \ N. \quad (27)$$

The SI values of Eqs. 26 and 27 equal each other, which validates the correctness of the quantum equation in SE units.

F. Ampère's magnetic force

In the 1820s, André Ampère[13] discovered that two parallel linear conductors, a distance r from each other, with electron current (i_1 and i_2) flowing, respectively, through each one, create, over a distance d, a magnetic force $\boldsymbol{F}_m$ between themselves.

1. Magnetic-force proportion

Ampère's magnetic-force proportion possesses an action of dimensions: charge squared per time squared $Q^2 \ T^{-2}$ and a reaction of the dimension: magnetic force $\boldsymbol{F}_m$ such that $\boldsymbol{F}_m \propto i_1 \ i_2 \ d \ r^{-1}$, which expands into $\boldsymbol{F}_m \propto q_1{\cdot}s^{-1} \ q_2{\cdot}s^{-1} \ d \ r^{-1}$. Later, the proportion was revised to $\boldsymbol{F}_m \propto q_1{\cdot}s^{-1} \ q_2{\cdot}s^{-1} \ d \ (2 \ \pi \ r)^{-1}$, where $(2 \ \pi \ r)$ is the length of the circumference of the imaginary cylinder (of length d), which is concentric

to one of the linear conductors. The other linear conductor resides longitudi-
nally on that cylinder's surface.

2. Permeability constant in SI units

The permeability constant[14] converts the magnetic-force proportion into the
magnetic-force equation. Unlike the process for determining the value of the
permittivity constant by establishing a proportion (see Section III.E.2.), the
permeability constant (*NIST* calls this constant the *magnetic constant*) , in
the emu system of unit measures, was *assigned* a value of unity, which,
when converted to the SI value, is (see Table VI):

$$\mu_0 = 4 \pi \times 10^{-7} \text{ N·C}^{-2}\text{·s}^2. \tag{28}$$

3. Magnetic-force equation in SI units

A magnetic-force equation was used to *define* the magnitude of the SI unit
of electric current, the ampère A (for short: amp), which was named after
the man who discovered this phenomenon. To obtain the magnitude and SI
value of the ampère A, the SI magnetic-force equation used is:

$$\boldsymbol{F}_{mm} = \mu_0 \; i_1 \; i_2 \; d \; (2 \pi r)^{-1}, \tag{29}$$

where:

$\boldsymbol{F}_{mm} = 2 \times 10^{-7}$ newton of magnetic force,
$\mu_0 = 4 \pi \times 10^{-7}$ henry per meter (or $\text{N·C}^{-2}\text{·s}^2$), the permeability
        constant,
$i_1 = i_2$ in ampere ($i_1 \; i_2 = i^2$), and
$d = r = 1$ meter,

thus:

$$(2 \times 10^{-7} \text{ N}) = (4 \pi \times 10^{-7} \text{ N·C}^{-2}\text{·s}^2)$$
$$\times i^2 \; (1 \text{ m}) \; [2 \pi (1 \text{ m})]^{-1} \tag{30}$$

and, when cross cancelled and simplified, gives $i^2 = 1 \text{ A}^2$; therefore,
$i_1 = i_2 = 1$ A. By definition, $1 \text{ A} = 1 \text{ C·s}^{-1}$ and, because the magnitude of s is already defined, defining that of A actually defines that of C
also.

4. Permeability constant in SE units

Substituting SE units for the SI units gives:

$$\mu_0 = (4 \pi \times 10^{-7}) \; [(3.374 \times 10^{-2})^{-1} \; \boldsymbol{f}_{ie}]$$
$$\times \; [(1.602 \times 10^{-19})^{-1} \; q_e)]^{-2} \; [(8.093 \times 10^{-21})^{-1} \; t_e]^2$$
$$= (68.518)^{-1} \; \boldsymbol{f}_{ie} \cdot q_e^{-2} \cdot t_e^2 = 2 \; \alpha \; \boldsymbol{f}_{ie} \cdot q_e^{-2} \cdot t_e^2. \qquad (31)$$

The $(68.518)^{-1}$ value is twice the value of the fine-structure constant. As with the permittivity constant earlier, we *also* realize that the permeability constant is not really a fundamental constant either.

5. Magnetic-force equation in SE units

Separating the values from the units in the SI magnetic-force equation (see Eq. 29), in preparation for creating the magnetic-force equation in SE units, gives:

$$\boldsymbol{F}_{mm} = (4 \pi \times 10^{-7} \; \text{N} \cdot \text{C}^{-2} \cdot \text{s}^2) \; (|i_1|_m \; \text{C} \cdot \text{s}^{-1}) \; (|i_2|_m \; \text{C} \cdot \text{s}^{-1})$$
$$\times \; (|d|_m \; \text{m}) \; (2 \pi \; |r|_m \; \text{m})^{-1}. \qquad (32)$$

Replacing the SI units with their SE units of the same magnitude gives:

$$\boldsymbol{F}_{me} = (2 \; \alpha \; \boldsymbol{f}_{ie} \cdot q_e^{-2} \cdot t_e^2) \; (|i_1|_e \; q_e \cdot t_e^{-1}) \; (|i_2|_e \; q_e \cdot t_e^{-1})$$
$$\times \; (|d|_e \; \lambda_e) \; (2 \pi \; |r|_e \; \lambda_e)^{-1}. \qquad (33)$$

Cross cancelling units and simplifying gives:

$$\boldsymbol{F}_{me} = 2 \; |i_1|_e \; |i_2|_e \; |d|_e \; (2 \pi \; |r|_e)^{-1} \; (\alpha \; \boldsymbol{f}_{ie}). \qquad (34)$$

The factors $(\alpha \; \boldsymbol{f}_{ie})$ are equal to the factor $\boldsymbol{f}_{me}$ (see Table VIII), which is the SE unit of magnetic force, because, historically, the value of the permeability constant reflected the practice of using an inertial-force-based unit of force in defining its value rather than a magnetic-based force. This means that the *correct* SE magnetic-force equation is (see Eq. 39 and the section that follows this one):

$$\boldsymbol{F}_{me} = 2 \; |i_1|_e \; |i_2|_e \; |d|_e \; (2 \pi \; |r|_e)^{-1} \; \boldsymbol{f}_{me}. \qquad (35)$$

6. SE unit of magnetic force

The SE unit of magnetic force $\boldsymbol{f}_{me}$ is the reaction to one electron separated from another one by the SE unit of length λ_e where each electron flows parallel to the other over a distance of λ_e such that $\boldsymbol{f}_{me}$ is proportional to, but not equal to, $q_e^2 \cdot t_e^{-2}$. The action is $q_e^2 \cdot t_e^{-2}$, the reaction, $\boldsymbol{f}_{me}$. All other possible magnitudes are positive, integral multiples of this magnetic-force quantum. Its proportion is:

$$\boldsymbol{f}_{me} \propto q_e^2 \cdot t_e^{-2}. \tag{36}$$

Using Ampère's historical SI magnetic-force equation, which includes the permeability constant as a factor (see Section III.F.4.), to calculate the magnitude of $\boldsymbol{f}_{me}$ gives a *preliminary* SI value of:

$$\begin{aligned}
\boldsymbol{f}_{mem} &= \mu_0 \ (q_e \cdot t_e^{-1})^2 \ (\lambda_e) \ (2 \ \pi \ \lambda_e)^{-1} = (4 \ \pi \times 10^{-7} \ \mathrm{N \cdot C^{-2} \cdot s^2}) \\
&\times \ [(1.602 \times 10^{-19} \ \mathrm{C}) \ (8.093 \times 10^{-21} \ \mathrm{s})^{-1}]^2 \ (2 \ \pi)^{-1} \\
&= 7.838 \times 10^{-5} \ \mathrm{N}.
\end{aligned} \tag{37}$$

At first, the quantum magnetic-force equation appears to be:

$$\boldsymbol{F}_{me} = |i_1|_e \ |i_2|_e \ |d|_e \ |r|_e^{-1} \ \boldsymbol{f}_{me}, \tag{38}$$

where:

$\quad |d|_e$ = an integral number of SE units of length of each of two conduc-
$\qquad$ tors over which the force is measured,

$\quad |i_1|_e$ and $|i_2|_e$ each = an integral number of electrons per the SE unit of
$\qquad$ time, and

$\quad |r|_e$ = an integral number of SE units of length between the two conduc-
$\qquad$ tors.

However, cylindrical symmetry *requires* that the quantum equation be:

$$\boldsymbol{F}_{me} = 2 \ |i_1|_e \ |i_2|_e \ |d|_e \ (2 \ \pi \ |r|_e)^{-1} \ \boldsymbol{f}_{me}, \tag{39}$$

which places 2 and $(2 \ \pi \ r)$, the length of the cylinder's circumference (of points of equal force), into the equation. Therefore, the *correct* magnitude of $\boldsymbol{f}_{me}$ in the SI unit of force is (see Eq. 37 and Table VIII):

$$\boldsymbol{f}_{mem} = \pi \ (7.838 \times 10^{-5} \ \mathrm{N}) = 2.462 \times 10^{-4} \ \mathrm{N}. \tag{40}$$

Comparing Eq. 23 with Eq. 40 shows that the SI values of the electronic-force quantum f_{eem} and the magnetic-force quantum f_{mem} *are equal to each other*. Wow! Just this comparison shows that both of the electronic and magnetic forces are different aspects of the *same* force.

Comparing Eq. 4 with Eq. 40 shows that the magnitude of the quantum of electron-based inertial force is $137.\text{...}$ times greater than that of magnetic force (see Table VIII):

$$f_{iem} \ f_{mem}^{-1} = (3.374 \times 10^{-2} \ \text{N}) \ (2.462 \times 10^{-4} \ \text{N})^{-1}$$
$$= 137.\text{...} \tag{41}$$

This magnetic phenomenon gives one definition of the fine-structure constant. The electronic- and gravitational-force phenomena give two more. See Eqs. 24 and 60.

7. Comparing quantum and classical magnetic-force equations

In an example quantum magnetic-force equation, let:

$$|i_1|_e = |q_1|_e \ |t_1|_e^{-1} = (3.5 \times 10^{12}) \ (1.8 \times 10^6)^{-1},$$
$$|i_2|_e = |q_2|_e \ |t_2|_e^{-1} = (2.0 \times 10^{10}) \ (1.8 \times 10^6)^{-1},$$
$$t_1 = t_2, \text{ and}$$
$d \ - \ r$, so that, in the equation, they cancel each other.

These magnitudes, using SI values, are:

$$q_1 = (3.5 \times 10^{12}) \ (1.602 \times 10^{-19} \ \text{C}) = 5.607 \times 10^{-7} \ \text{C},$$
$$q_2 = (2.0 \times 10^{10}) \ (1.602 \times 10^{-19} \ \text{C}) = 3.204 \times 10^{-9} \ \text{C},$$
$$t_1 = t_2 = (1.8 \times 10^6) \ (8.093 \times 10^{-21} \ \text{s})$$
$$= 1.456 \times 10^{-14} \ \text{s}.$$

The quantum equation is:

$$F_{me} = 2 \ [(3.5 \times 10^{12}) \ (1.8 \times 10^6)^{-1}]$$
$$\times \ [(2.0 \times 10^{10}) \ (1.8 \times 10^6)^{-1}] \ (2 \ \pi)^{-1} \ f_{me}$$
$$= 6.877 \times 10^9 \ f_{me}. \tag{42}$$

For later comparison, replacing f_{me} with its SI value gives (see Eq. 40):

$$F_{mem} = (6.877 \times 10^9) \ (2.462 \times 10^{-4} \ \text{N})$$
$$= 1.692 \times 10^6 \ \text{N}. \tag{43}$$

The classical equation is:

$$
\begin{aligned}
\boldsymbol{F}_{mm} &= \mu_0 \; i_1 \; i_2 \; d \; (2 \; \pi \; r)^{-1} = (4 \; \pi \times 10^{-7} \; \text{N·C}^{-2}\text{·s}^2) \\
&\times [(5.607 \times 10^{-7} \; \text{C}) \; (1.456 \times 10^{-14} \; \text{s})^{-1}] \\
&\times [(3.204 \times 10^{-9} \; \text{C}) \; (1.456 \times 10^{-14} \; \text{s})^{-1}] \\
&\times (2 \; \pi)^{-1} = 1.692 \times 10^6 \; \text{N}, \quad\quad (44)
\end{aligned}
$$

where the value of d cancels that of r. The SI values of Eqs. 43 and 44 equal each other, which validates the correctness of the quantum equation in SE units.

8. Mathematical product of ϵ_0 and μ_0

The product[15] of ϵ_0 and μ_0 is:

$$
\begin{aligned}
\epsilon_0 \; \mu_0 &= [(2 \; \alpha)^{-1} \; \boldsymbol{f}_{ie}^{-1} \cdot q_e^2 \cdot \lambda_e^{-2}] \; [2 \; \alpha \; \boldsymbol{f}_{ie} \cdot q_e^{-2} \cdot t_e^2] \\
&= \lambda_e^{-2} \cdot t_e^2 = c^{-2},
\end{aligned}
$$

or

$$
c = (\epsilon_0 \; \mu_0)^{-\frac{1}{2}}, \quad\quad (45)
$$

the speed of light.

G. Newton's gravitational force

In the 1680s, Isaac Newton[16] discovered that two masses (m_1 and m_2) a distance r from each other create a gravitational force of attraction $\boldsymbol{F}_g$ between themselves.

1. Gravitational-force proportion

Newton's gravitational-force proportion possesses an action of dimensions: mass squared per length squared $M^2 L^{-2}$ and a reaction of the dimension: gravitational force $\boldsymbol{F}_g$ such that $\boldsymbol{F}_g \propto m_1 \; m_2 \; r^{-2}$.

2. Gravitation constant in SI units

The gravitation constant[17] G converts the gravitational-force proportion into the gravitational-force equation. An example experiment shows how the SI value of the gravitation constant was determined (Newton did not determine this SI value because, at the time, the metric system did not yet exist): Let the action arbitrarily be set up as a point mass of $m_1 = 6.0 \times 10^{24}$ kg separated from a second point mass of $m_2 = 1.0 \times 10^3$ kg by a distance of $r = 6.4 \times 10^6$ m. In this experiment, actually measuring the gravitational force of attraction between the two masses would yield a reaction of $\pmb{F}_{gm} = 9.776 \times 10^3$ N. The proportion is:

$$9.776 \times 10^3 \text{ N} \propto (6.0 \times 10^{24} \text{ kg})$$
$$\times (1.0 \times 10^3 \text{ kg}) (6.4 \times 10^6 \text{ m})^{-2} \qquad (46)$$
$$\text{or}$$
$$9.776 \times 10^3 \text{ N} \propto 1.464 \times 10^{14} \text{ kg}^2 \cdot \text{m}^{-2}. \qquad (47)$$

By changing the magnitudes of the action's three arguments, the action's value would change by a particular percentage, and the reaction's value would change by that same percentage; therefore, no matter the values of the action's three factors, dividing the value of the reaction by that of the action yields the same value—a constant of proportionality—the gravitation constant, which *NIST* calls the *Newtonian constant of gravitation* (see Table VII):

$$G = (9.776 \times 10^3 \text{ N}) (1.464 \times 10^{14} \text{ kg}^2 \cdot \text{m}^{-2})^{-1}$$
$$= 6.674 \times 10^{-11} \text{ N} \cdot \text{kg}^{-2} \cdot \text{m}^2. \qquad (48)$$

3. Gravitational-force equation in SI units

We already know that the SI gravitational-force equation is:

$$\pmb{F}_{gm} = G \, m_1 \, m_2 \, r^{-2}, \qquad (49)$$

where:
   m_1 and m_2 each represent a mass in kilograms,
   r is the distance between them in meters, and
   G is the gravitation constant in $\text{kg}^{-1} \cdot \text{m}^3 \cdot \text{s}^{-2}$ or in $\text{N} \cdot \text{kg}^{-2} \cdot \text{m}^2$ where
   $N = \text{kg} \cdot \text{m} \cdot \text{s}^{-2}$.

4. Gravitation constant in SG units

To determine the value of the gravitation constant G in SG units, substitute SG units for the SI units, which gives:

$$G = (6.674 \times 10^{-11}) \; [(1.405 \times 10^{41})^{-1} \, \pmb{f}_{ig}]$$
$$\times \; [(1.188 \times 10^{-33})^{-1} \, \lambda_g]^2 \; \times \; [(1.859 \times 10^{-9})^{-1} \, m_g]^{-2}$$
$$= 1.161 \times 10^{-3} \, \pmb{f}_{ig} \cdot \lambda_g^2 \cdot m_g^{-2}$$
$$= 2 \, \alpha \, (4 \, \pi)^{-1} \, \pmb{f}_{ig} \cdot \lambda_g^2 \cdot m_g^{-2}. \qquad (50)$$

The value, 1.161×10^{-3}, is twice the value of the fine-structure constant divided by $(4\ \pi)$. In the gravitational-force equation, the $(4\ \pi)$ value is a factor in the formula for the area $(4\ \pi\ r^2)$ of the imaginary sphere's surface of equal force, but, historically, it was placed in the G constant. Here again, like the permittivity and permeability constants, earlier, the gravitation constant is shown not to be a fundamental constant either.

5. Gravitational-force equation in SG units

Separating the values from the units in the SI gravitational-force equation (see Eq. 49), in preparation for creating the electronic-force equation in SG units, gives:

$$\mathbf{F}_{gm} = (6.674 \times 10^{-11} \ \text{N} \cdot \text{kg}^{-2} \cdot \text{m}^2)$$
$$\times \ (|m_1|_m \ \text{kg}) \ (|m_2|_m \ \text{kg}) \ (|r^2|_m \ \text{m}^2)^{-1}. \tag{51}$$

Replacing the SI units with their SG units of the same magnitude gives:

$$\mathbf{F}_{gg} = [2 \ \alpha \ (4\ \pi)^{-1} \ \mathbf{f}_{ig} \cdot \lambda_g^2 \cdot m_g^{-2}]$$
$$(|m_1|_g \ m_g) \ (|m_2|_g \ m_g) \ (|r^2|_g \ \lambda_g^2)^{-1}. \tag{52}$$

Note that $(|m_1|_m \ \text{kg}) = (|m_1|_g \ m_g)$ and $(|r^2|_m \ \text{m}^2) = (|r^2|_g \ \lambda_g^2)$.

Cross cancelling units and simplifying gives:

$$\mathbf{F}_{gg} = 2 \ |m_1|_g \ |m_2|_g \ (4\ \pi \ |r^2|_g)^{-1} \ (\alpha \ \mathbf{f}_{ig}). \tag{53}$$

The factors $(\alpha \ \mathbf{f}_{ig})$ are equal to the factor $\mathbf{f}_{gg}$ (see Table VIII), which is the SG unit of gravitational force, because, historically, the value of the gravitation constant reflected the practice of using an inertial-force-based unit of force in defining its value rather than a gravitational-force-based one. This means that the correct SG gravitational-force equation is (see Eq. 59 and the section that follows this one):

$$\mathbf{F}_{gg} = 2 \ |m_1|_g \ |m_2|_g \ (4\ \pi \ |r^2|_g)^{-1} \ \mathbf{f}_{gg}. \tag{54}$$

6. SG unit of gravitational force

The SG unit of gravitational force $\boldsymbol{f}_{gg}$ is the reaction to one masstron m_g separated from another masstron m_g by the SG unit of length λ_g such that $\boldsymbol{f}_{gg}$ is proportional to, but not equal to, $m_g{}^2 \cdot \lambda_g{}^{-2}$. The action is $m_g{}^2 \cdot \lambda_g{}^{-2}$, the reaction, $\boldsymbol{f}_{gg}$. All other possible magnitudes are positive, integral multiples of this gravitational-force quantum. Its proportion is:

$$\boldsymbol{f}_{gg} \propto m_g{}^2 \cdot \lambda_g{}^{-2}. \tag{55}$$

Using Newton's historical SI gravitational-force equation (see Section III.G.3.), which includes the gravitation constant as a factor, to calculate the magnitude of $\boldsymbol{f}_{gg}$ gives a *preliminary* SI value of:

$$
\begin{aligned}
\boldsymbol{f}_{ggm} &= G\, m_g{}^2\, \lambda_g{}^{-2} = (6.674 \times 10^{-11}\ \mathrm{N \cdot kg^{-2} \cdot m^2}) \\
&\times (1.859 \times 10^{-9}\ \mathrm{kg})^2\, (1.188 \times 10^{-33}\ \mathrm{m})^{-2} \\
&= 1.633 \times 10^{38}\ \mathrm{N}.
\end{aligned} \tag{56}
$$

At first, the quantum gravitational-force equation appears to be:

$$\boldsymbol{F}_{gg} = |m_1|_g\ |m_2|_g\ |r|_g{}^{-2}\ \boldsymbol{f}_{gg}, \tag{57}$$

where:

   $|m_1|_g$ and $|m_2|_g$ each represent an integral number of masstrons, and $|r|_g$ is the integral number of SG units of length between them.

However, spherical symmetry *requires* that the equation be:

$$\boldsymbol{F}_{gg} = 2\ |m_1|_g\ |m_2|_g\ (4\ \pi\ |r|_g{}^2)^{-1}\ \boldsymbol{f}_{gg}, \tag{58}$$

which places 2 and $(4\ \pi\ r^2)$, the area of the surface of the sphere of equal force, into the equation. Therefore, the *correct* magnitude of $\boldsymbol{f}_{gg}$ in the SI unit of force is (see Eq. 56 and Table VIII):

$$
\begin{aligned}
\boldsymbol{f}_{ggm} &= \tfrac{1}{2}\ (4\ \pi)\ (1.633 \times 10^{38}\ \mathrm{N}) \\
&= 1.026 \times 10^{39}\ \mathrm{N}.
\end{aligned} \tag{59}
$$

Comparing Eq. 5 with Eq. 59 shows that the magnitude of the quantum of gravitational-based inertial force is $137 . \dots$ times greater than that of gravitational force (see Table VIII):

$$\boldsymbol{f}_{igm}\ \boldsymbol{f}_{ggm}^{-1} = (1.405 \times 10^{41}\ N)\ (1.026 \times 10^{39}\ N)^{-1}$$
$$= 137.... \tag{60}$$

This gravitational phenomenon gives one definition of the fine-structure constant. The electronic- and magnetic-force phenomena give two more. See Eqs. 24 and 41.

7. Comparing quantum and classical gravitational-force equations

In an example quantum gravitational-force equation, let:

$|m_1|_g = 6.2 \times 10^{45}$ (masstrons),
$|m_2|_g = 1.8 \times 10^{20}$ (masstrons), and
$|r|_g = 8.3 \times 10^{65}$ (quantum gravitational units of length).

These magnitudes, using SI values, are:

$m_1 = (6.2 \times 10^{45})\ (1.859 \times 10^{-9}\ kg) = 1.152 \times 10^{37}\ kg,$
$m_2 = (1.8 \times 10^{20})\ (1.859 \times 10^{-9}\ kg) = 3.346 \times 10^{11}\ kg,$
and
$r = (8.3 \times 10^{65})\ (1.188 \times 10^{-33}\ m) = 9.860 \times 10^{32}\ m.$

The quantum equation in SG units is:

$$F_{gg} = 2\ (6.2 \times 10^{45})\ (1.8 \times 10^{20})$$
$$\times\ [4\ \pi\ (8.3 \times 10^{65})^2]^{-1}\ \boldsymbol{f}_{gg} = 2.578 \times 10^{-67}\ \boldsymbol{f}_{gg}. \tag{61}$$

For later comparison, replacing $\boldsymbol{f}_{gg}$ with its SI value (see Eq. 59) gives:

$$F_{ggm} = (2.578 \times 10^{-67})\ (1.026 \times 10^{39}\ N)$$
$$= 2.643 \times 10^{-28}\ N. \tag{62}$$

The classical equation in SI units is:

$$F_{gm} = (6.674 \times 10^{-11}\ N \cdot kg^{-2} \cdot m^2)\ (1.152 \times 10^{37}\ kg)$$
$$\times\ (3.346 \times 10^{11}\ kg)\ (9.860 \times 10^{32}\ m)^{-2}$$
$$= 2.643 \times 10^{-28}\ N. \tag{63}$$

The SI values of Eqs. 62 and 63 equal each other, which validates the correctness of the quantum equation in SG units.

8. Planck values

A comparison between the historical Planck values [18] and the quantum attributes of the masstron reveals differences between the two as shown in Table V. The Planck-mass and -temperature calculations use the $G^{-\frac{1}{2}}$ factor, and, because G contains the α factor, they are equal, respectively, to $(\alpha^{-\frac{1}{2}} m_g)$ and $(\alpha^{-\frac{1}{2}} k_g)$. The Planck-length and -time calculations use the $G^{\frac{1}{2}}$ factor, and, because G contains the α factor, they are equal, respectively, to $[\alpha^{\frac{1}{2}} (2\pi)^{-1} \lambda_g]$ and $[\alpha^{\frac{1}{2}} (2\pi)^{-1} t_g]$. The $[(2\pi)^{-1}]$ factor occurs because $\hbar$ and G, together, contain $[(4\pi^2)^{-1}]$. The attributes of the masstron *now* appear to be the *correct* values for the Planck values.

9. Discussion

Each of the permittivity-, permeability-, and gravitational-force constants contains the fine-structure constant as one of its factors (see, respectively, Eqs. 13, 30, and 50). This occurs because the forces that are involved in these constants' respective equations were historically measured using the SI unit (newton) of inertial force and not a quantum-electromagnetic or -gravitational unit of force, in which case, the fine-structure constant would vanish. See Tables VI through X. *This seems to mean that the magnitude of the fine-structure constant is inversely proportional to that of inertial force.*

H. Sommerfeld's fine-structure constant

The historical SI definition of Arnold Sommerfeld's fine-structure constant [19] is:

$$\alpha = 2 \pi e^2 (h c)^{-1}. \tag{64}$$

This section shows that α is *not* related to the e^2, h, and c factors, which, historically, have been used to define it. Those three factors are composed of more-basic factors, which cancel each other within the definition of α except for α itself, which is a factor within e^2 or, more precisely, within ϵ_0. Converting the SI factors into SE factors gives:

$$e^2 = q_e^2 (4\pi \epsilon_0)^{-1},$$
$$\epsilon_0 = q_e^2 (2\alpha E_e \cdot \lambda_e)^{-1},$$
$$E_e = m_e \cdot \lambda_e^2 \cdot t_e^{-2},$$
$$h = E_e \cdot t_e, \text{ and}$$
$$c = \lambda_e \cdot t_e^{-1}.$$

Using these SE factors for α gives:

$$\alpha = 2 \pi q_e^2 \{ 4 \pi [q_e^2 (2 \alpha E_e \cdot \lambda_e)^{-1}] \}^{-1}$$
$$\times [(E_e \cdot t_e) (\lambda_e \cdot t_e^{-1})]^{-1}. \tag{65}$$

Collecting factors into one numerator and one denominator gives:

$$\alpha = (2 \pi q_e^2 \ 2 \alpha E_e \cdot \lambda_e \cdot t_e) (4 \pi q_e^2 \cdot E_e \cdot t_e \cdot \lambda_e)^{-1}. \tag{66}$$

Cross cancelling matching factors gives:

$$\alpha = \alpha. \tag{67}$$

All of the numerator factors cross cancel the denominator ones except for α. Therefore, the fine-structure constant is more basic than the fundamental constants that, historically, have been used to define it. In sum, this constant was defined recursively—in terms of itself.

1. Discussion

The fine-structure constant is the ratio between the magnitude of the quantum unit of electromagnetic force and a magnitude of inertial force that uses the quantum attributes of the electron for its mass and acceleration. See Eqs. 24 and 41 and Table VIII.

The fine-structure constant is also the ratio between the magnitude of the quantum unit of gravitational force and a magnitude of inertial force that uses the quantum attributes of the masstron for its mass and acceleration. See Eq. 60 and Table IX.

The meaning of the magnitude of the fine-structure constant appears to be the ratio between the minimum magnitude of inertial force in the Universe and the local (in the vicinity of the Solar System) magnitude of inertial force. For example, perhaps a place in the Universe exists where a one-kilogram mass can be accelerated at a rate of one meter per second per second by applying only one-137th newton of force rather than a full newton, which is required in the vicinity of the Solar System. In essence, the magnitude of the fine-structure constant seems to be systemic (local), not universal (see Section III. in *Part I: The Voyage*).

Apparently, numerologists, who, over the years, have tried to discover a purely-mathematical formula for the definition of the fine-structure constant, have (in my humble opinion) wasted their time—there is none.

I. Bohr's hydrogen atom

All of the SE attributes of the bound electron[20] in Niels Bohr's model of Ernest Rutherford's hydrogen atom contain *only one constant*: the fine-structure constant α (besides π for angular attributes). See Table XI. For example, the historical SI formula[21] for the orbit radius a_n is:

$$a_n = \epsilon_0 \ h^2 \ n^2 \ (\pi \ m_e \cdot q_e^2)^{-1}. \tag{68}$$

It contains the M and Q dimensions in addition to all of those dimensions in the permittivity and Planck's constants. By converting these two constants into their SE values and simplifying gives the SE orbit radius in its proper L dimension:

$$a_n = n^2 \ (m_e \cdot \lambda_e^2 \cdot t_e^{-1})^2 \ [\,(2 \ \alpha)^{-1} \ t_e^2 \cdot q_e^2 \cdot m_e^{-1} \cdot \lambda_e^{-3}\,] \ (\pi \ m_e \cdot q_e^2)^{-1}$$
$$= n^2 \ (2 \ \pi \ \alpha)^{-1} \ \lambda_e. \tag{69}$$

Multiplying by $(2 \ \pi)$ gives the orbit length, which, in the ground state $(n = 1)$, is about $137\ldots$ times λ_e. The other attributes of the bound electron are converted into SE values in the same way (see Table XI). For the SE values for hydrogen energy and photon emission, see Tables XII and XIII, respectively.

1. Bohr units of length and time

Two factors affect the bound electron that do not affect free electrons. The disparity between the magnitudes of the SE inertial $\boldsymbol{f}_{ie}$ and electromagnetic $\boldsymbol{f}_{ee}$ quantum units of force appears to introduce the fine-structure constant α (see Table XI). Quantum discontinuity introduces the orbit number n.

Whenever an equation, which deals with the Bohr model of the hydrogen atom, contains the SE unit of length λ_e or time t_e or the fine-structure constant α, the orbit number is included as a factor. For example: $(n \ \lambda_e)$, $(n \ t_e)$, and $(n \ \alpha^{-1})$. In any orbit, the variation of mass m_e of the bound electron is negligible; therefore, n is not included as a factor. See Table XI under the heading of the column labeled: **Quantum attribute-factor configuration**.

Defining the Bohr units of length λ_n and time t_n as:

$$\lambda_n = (n \ \alpha^{-1}) \ (n \ \lambda_e) = n^2 \ \alpha^{-1} \ \lambda_e, \tag{70}$$

and

$$t_n = (n \ \alpha^{-1})^2 \ (n \ t_e) = n^3 \ \alpha^{-2} \ t_e, \tag{71}$$

and expressing the remaining attributes using these Bohr units simplifies the definitions even more, as seen in Table XI under the heading of the column labeled: **Code (in Bohr units)**.

2. Discussion

The fine-structure constant α is a factor in every one of the attributes of the bound electron—*to the exclusion of any other constant* (except the geometrical constant π). Although the orbit angular momentum appears not to contain the fine-structure constant, it is composed of a numerator and denominator, both of which contain it as a factor; therefore, these α factors cross cancel each other (see Table XI).

   The fine-structure constant seems to be present because both the electromagnetic and inertial forces act upon the bound electron. If the fine-structure constant is a local (rather than a universal) constant, perhaps at other locations in the Universe, electronic orbits in atoms can be either closer to the nucleus or farther away than their locations in atoms in the vicinity of the Solar System. For example, at a place in the Universe where α^{-1} equals 140, the Bohr radius would be $[140 \ (2 \ \pi)^{-1} \ \lambda_e]$ rather than $[137 \ (2 \ \pi)^{-1} \ \lambda_e]$, which is the Solar-System value. If this be the case, the larger quantum magnitude of inertial force would pull the electron orbits farther away from the nucleus of an atom (see Section III. in *Part I: The Voyage*).

J. Planck's cavity radiancy

Max Planck's SI equation for cavity radiancy,[22] which calculates the amount of energy $\mathbb{R}_\Lambda$ of a particular wavelength Λ that is radiated from a cavity at a particular temperature K is:

$$\mathbb{R}_\Lambda = c_1 \, \Lambda^{-5} \, [\exp(c_2 \, K^{-1} \, \Lambda^{-1}) - 1]^{-1}, \tag{72}$$

where:

$c_1 = 2 \pi c^2 h = 3.741 \times 10^{-16}$ kg·m⁴·s⁻³ or J·s⁻¹·m²,

$c_2 = h c k^{-1} = 1.438 \times 10^{-2}$ K·m,

K = the temperature of the cavity in kelvin,

Λ = the wavelength of the radiancy in meters, and

$\exp(x)$ = e, which is $2.71828...$, to the power of x where,

$x = c_2 \, K^{-1} \, \Lambda^{-1}$, the exponent of base e.

Expressing the two Wilhelm Wien radiation constants,[23] c_1 and c_2, with SE units gives (see Table VI):

$$c_1 = 2 \pi c^2 h = 2 \pi (\lambda_e \cdot t_e^{-1})^2 (m_e \cdot \lambda_e^2 \cdot t_e^{-1})$$
$$= (E_e \cdot t_e^{-1}) (2 \pi \lambda_e^2), \tag{73}$$

which now can be read as: c_1 is the electron's rest-mass energy E_e, which radiates during each t_e quantum time period from a $(2 \pi \lambda_e^2)$ quantum hemisphere surface area.

$$c_2 = h c k^{-1} = (m_e \cdot \lambda_e^2 \cdot t_e^{-1}) (\lambda_e \cdot t_e^{-1}) (m_e \cdot \lambda_e^2 \cdot t_e^{-2} \cdot k_e^{-1})^{-1}$$
$$= k_e \cdot \lambda_e. \tag{74}$$

See Table VI. Notice that both c_1 and c_2 contain Planck's constant. This is where Planck first observed two equal values with the same units—one in each of Wien's constants. Later, that value was named Planck's constant.

The Planck radiancy equation, using SE units, is:

$$\mathbb{R}_\Lambda = 2 \pi \Lambda^{-5} \, [\exp(K^{-1} \, \Lambda^{-1}) - 1]^{-1}, \tag{75}$$

where the magnitudes of the temperature K and wavelength Λ variables are in SE units.

1. Comparing SI and SE radiancy equations

The following example problem calculates the magnitude of the radiancy at wavelength Λ from a cavity at temperature K where the SI and SE values for the magnitudes of Λ and K are, respectively, as follows:

$$\Lambda = 1.5 \times 10^{-6} \text{ m} = (1.5 \times 10^{-6}) \; [(2.426 \times 10^{-12})^{-1} \; \lambda_e]$$
$$= 6.182 \times 10^5 \; \lambda_e. \tag{76}$$

and

$$K = 2000 \text{ K} = (2000) \; [(5.929 \times 10^{9})^{-1} \; k_e]$$
$$= 3.372 \times 10^{-7} \; k_e. \tag{77}$$

Using Planck's equation with the empirical SI values for c_1 and c_2 (see Table VI):

$$R_\Lambda = c_1 \; \Lambda^{-5} \; [\exp(c_2 \; K^{-1} \; \Lambda^{-1}) - 1]^{-1}. \tag{78}$$

$$R_\Lambda = (3.741 \times 10^{-16} \text{ J·s}^{-1}\text{·m}^2) \; (1.5 \times 10^{-6} \text{ m})^{-5}$$
$$\times \{\exp[(1.438 \times 10^{-2} \text{ K·m}) \; (2000 \text{ K})^{-1}$$
$$\times (1.5 \times 10^{-6} \text{ m})^{-1}] - 1\}^{-1}$$
$$= 4.105 \times 10^{11} \text{ J·s}^{-1}\text{·m}^{-3}. \tag{79}$$

The SE equation is (see Eqs. 76 and 77):

$$R_\Lambda = 2 \; \pi \; \Lambda^{-5} \; [\exp(K^{-1} \; \Lambda^{-1}) - 1]^{-1}. \tag{80}$$

$$R_\Lambda = 2 \; \pi \; (6.182 \times 10^5 \; \lambda_e)^{-5}$$
$$\times \{\exp[(3.372 \times 10^{-7} \; k_e)^{-1} \; (6.182 \times 10^5 \; \lambda_e)^{-1}] - 1\}^{-1}$$
$$= 5.797 \times 10^{-31} \text{ E}_e\text{·t}_e^{-1}\text{·}\lambda_e^{-3}. \tag{81}$$

To compare the SI and SE solutions, the SE units are converted into SI units:

$$R_\Lambda = (5.791 \times 10^{-31}) \; (8.187 \times 10^{-14} \text{ J})$$
$$\times (8.093 \times 10^{-21} \text{ s})^{-1} \; (2.426 \times 10^{-12} \text{ m})^{-3}$$
$$= 4.105 \times 10^{11} \text{ J·s}^{-1}\text{·m}^{-3}. \tag{82}$$

The SI values of Eqs. 79 and 82 are equal to each other.

2. Discussion

When using a system of unit measures based upon the quantum attributes of the electron, Wilhelm Wien's c_1 and c_2 constants in Max Planck's black-body-radiancy equation vanish—except for π. See Table VI.

K. Fizeau's speed of light

Over a century ago, Armand Fizeau was the first person to determine the speed of light experimentally. Using SI units, the speed of light[24] in a vacuum is now *defined* to be exactly:

$$c = 299,792,458 \text{ m·s}^{-1}. \tag{83}$$

See Tables VI and VII. Substituting SE units (or SP or SG units) for the SI units, gives a value of one. For example:

$$c = (2.997 \times 10^8) \ [(2.426 \times 10^{-12})^{-1} \ \lambda_e]$$
$$\times \ [(8.093 \times 10^{-21})^{-1} \ t_e]^{-1} = 1 \ \lambda_e \cdot t_e^{-1}. \tag{84}$$

L. Einstein's quantum energies

Albert Einstein's SI energy equation for mass[25] is:

$$E_m = m_m \ c^2. \tag{85}$$

Using SE units, where $(|c|_e = 1)$, it is:

$$E_e = m_e. \tag{86}$$

Einstein's SI energy equation for photons[26] is:

$$E_{ph} = h \ \nu_{ph}. \tag{87}$$

Using SE units, $(\nu_{ph} = |\nu|_e \ t_e^{-1})$ and $(h = m_e \cdot \lambda_e^2 \cdot t_e^{-1})$, so the SE equation is:

$$E_{ph} = (m_e \cdot \lambda_e^2 \cdot t_e^{-1}) \ (|\nu|_e \ t_e^{-1}) = E_e \ |\nu|_e, \tag{88}$$

where:

$E_e = (m_e \cdot \lambda_e^2 \cdot t_e^{-2})$ and is the rest-mass energy of the electron, and

$|\nu|_e$ is a dimensionless whole number, which represents the number of photons contained within one t_e of the electromagnetic beam and, because the beam travels at the speed of light, within one λ_e, as well.

1. Discussion

Apparently, the amount of energy contained within an interval of one t_e (or one λ_e) of the electromagnetic beam is equal to the number of photons found in that interval times the rest-mass energy of the electron. This seems to indicate that the energy of the electron and of the photon are the same and prompts speculation that, perhaps, the photon is a form of the electron. Yet, the electron possesses mass, and the photon does not. However, after examining the known attributes of electrons, positrons, and photons, a duality-based model of the photon emerges. The success of this duality model of the photon depends upon the polarity of all of the attributes of matter being opposite to those of antimatter, *including mass*. Reasonably, an antimatter world can exist only when *all* of the attributes of its elementary particles are opposite in sign to those of a matter world.

This duality model of photons, in its simplest form, is, as follows:

- Photons are not fundamental entities. They are composed of equal quantities of interlocking electrons and positrons in some form of energy mode of existence (see Section IV.A.4.).
- Photons are massless. The mass of their electronic part cancels the antimass of their positronic part. This masslessness and the forces generated between the electrons and positrons enable (or force) them to make a series of quantum leaps in the same direction at the speed of light.
- Massless photons (and gravitons), which are composed of equal quantities of matter and antimatter, are the links between the matter and antimatter parts of the Universe (see Section IV.A.1.).

Speculatively, all particles *must* possess mass and occupy space. "Massless" particles, such as photons and gravitons, are, supposedly, composed of half matter and half antimatter. The matter half possesses mass and volume, and the antimatter half possesses antimass and antivolume. The sums of these attributes and antiattributes are "zero" such that "massless" particles indeed appear to be massless and to occupy no space; therefore, a truly massless particle cannot exist. Probably, this is the reason that electromagnetic beams do not interfere with each other as they crisscross among themselves without modifying their respective attributes.

The momentum equation for a massive particle is ($p = m\ v$), where m is the mass of the particle and v is its velocity. A photon has no mass, and the magnitude of its velocity is the speed of light c, so the classical momentum equation seems not to work with a photon. Louis de Broglie's quantum-momentum equation for a photon, which is related to Albert Einstein's photon quantum-energy equation, is ($p_{ph} = h\ \lambda_{ph}^{-1}$), where h is Planck's constant and λ_{ph} is the wavelength of the photon.

In SE units, ($h = E_e\ t_e$) and ($\lambda_{ph} = |\lambda|_e\ \lambda_e$), so $p_{ph} = h\ \lambda_{ph}^{-1} = (E_e\ t_e)\ (|\lambda|_e\ \lambda_e)^{-1} = (m_e\ c^2\ t_e)\ (|\lambda|_e\ \lambda_e)^{-1} = m_e\ c\ (\ |\lambda|_e)^{-1}$. This equation reads: The momentum of a photon is equal to the SE rest mass of the electron multiplied by the speed of light divided by the SE wavelength of the photon (without its unit of length). Therefore, Louis de Broglie's photon momentum equation can contain a mass and a velocity after all.

M. Ohm's law

Georg Ohm's law is ($V = I\ R$),[27] where V is electric potential, I is electric current, and R is electric resistance.

The electric-potential quantum[28] is:

$$V_e = E_e \cdot q_e^{-1} = m_e \cdot \lambda_e^2 \cdot t_e^{-2} \cdot q_e^{-1}. \tag{89}$$

The electric-current quantum[29] is:

$$i_e = q_e \cdot t_e^{-1}. \tag{90}$$

The electric-resistance quantum[30] is:

$$R_e = V_e\ i_e^{-1} = (E_e \cdot q_e^{-1})\ (q_e \cdot t_e^{-1})^{-1} = E_e \cdot t_e \cdot q_e^{-2}. \tag{91}$$

The ($E_e \cdot t_e$) factors are equal to Planck's constant h (see Eq. 3); therefore, the quantum electrical resistance in its historical form[31] is:

$$R_e = h\ q_e^{-2}. \tag{92}$$

The electric-potential quantum V_e in terms of current and resistance is:

$$V_e = m_e \cdot \lambda_e^2 \cdot t_e^{-2} \cdot q_e^{-1} = (q_e \cdot t_e^{-1})\ (E_e \cdot t_e \cdot q_e^{-2}). \tag{93}$$

Calculating the magnetic-flux quantum[32] from the electric-resistance quantum gives:

$$\Phi_e = R_e \; q_e = (E_e \cdot t_e \cdot q_e^{-2}) \; (q_e) = E_e \cdot t_e \cdot q_e^{-1} = h \; q_e^{-1}. \quad (94)$$

N. Boltzmann's constant

The SI value of Ludwig Boltzmann's constant[33] is:

$$k = R \; N_A^{-1} = 1.381 \times 10^{-23} \; J \cdot K^{-1}, \quad (95)$$

where: R is the universal gas constant, and N_A is Amedeo Avogadro's number (see Table VI).
Replacing SI units with SE units gives (see Tables VI and VII):

$$k = (1.381 \times 10^{-23}) \; [(8.187 \times 10^{-14})^{-1} \; E_e]$$
$$\times \; [(5.929 \times 10^9)^{-1} \; k_e]^{-1} = 1 \; E_e \cdot k_e^{-1}. \quad (96)$$

Converting from SE to SG units gives:

$$k = (1 \; E_e \cdot k_e^{-1}) \; (2.041 \times 10^{21} \; E_e \cdot E_g^{-1})^{-1}$$
$$\times \; (2.041 \times 10^{21} \; k_e \cdot k_g^{-1}) = 1 \; E_g \cdot k_g^{-1}. \quad (97)$$

This SG value for Boltzmann's constant is equal to the Planck unit of entropy s_{P1} which is used in calculating Bekenstein and Hawking's equation for the entropy of a Black Hole[34] (see Section III.P.14.).

1. Avogadro's number

The SI value of Amedeo Avogadro's number[35] is:

$$N_A = 6.022 \times 10^{26} \; atoms \cdot kmole^{-1} \; or \; amu \cdot kg^{-1} = 1, \quad (98)$$

where: one amu (atomic mass unit)[36] is one-twelfth of the mass of the carbon-12 isotope (see Table VI). Avogadro's number is a conversion factor from one unit of mass to another, so its ultimate magnitude is dimensionless unity.

2. Universal gas constant

The value of the universal gas constant[37] is:

$$R = 8.314 \times 10^3 \text{ J·K}^{-1}\text{·atoms·kmole}^{-1}, \tag{99}$$

where: atoms·kmole^{-1} is the amu-to-kg conversion factor of 6.022×10^{26} ; therefore, the SI value of the universal gas constant is:

$$
\begin{aligned}
R_m &= (8.314 \times 10^3)\ (6.022 \times 10^{26})^{-1} \text{ J·K}^{-1} \\
&= 1.381 \times 10^{-23} \text{ J·K}^{-1}.
\end{aligned} \tag{100}
$$

Replacing SI units with SE units gives:

$$
\begin{aligned}
R_e &= (1.381 \times 10^{-23})\ [(8.187 \times 10^{-14})^{-1}\ E_e] \\
&\times\ [(5.929 \times 10^9)^{-1}\ k_e]^{-1} = 1\ E_e\text{·}k_e^{-1}.
\end{aligned} \tag{101}
$$

3. Discussion

Using SE units, Boltzmann's constant, in relation to Avogadro's number and the universal gas constant, is:

$$k_e = R\ N_A^{-1} = (1\ E_e\text{·}k_e^{-1})\ (1) = 1\ E_e\text{·}k_e^{-1}. \tag{102}$$

O. Magnetons

This section examines and compares the formulas for the Bohr and nuclear magnetons.

1. Bohr magneton

The SI Bohr-magneton equation[38] is:

$$\mu_B = h\ (4\ \pi)^{-1}\ q_e\text{·}m_e^{-1} = 9.274 \times 10^{-24} \text{ A·m}^2 \text{ or } \text{J·T}^{-1}. \tag{103}$$

The SE Bohr-magneton equation is:

$$\mu_B = (m_e\text{·}\lambda_e^2\text{·}t_e^{-1})\ (4\ \pi)^{-1}\ q_e\text{·}m_e^{-1} = (4\ \pi)^{-1}\ q_e\text{·}t_e^{-1}\text{·}\lambda_e^2. \tag{104}$$

2. Nuclear magneton

The SI nuclear-magneton equation [39] is:

$$\mu_N = h \ (4 \ \pi)^{-1} \ q_p \cdot m_p^{-1} = 5.050 \times 10^{-27} \ \text{A·m}^2. \quad (105)$$

The SP nuclear-magneton equation is:

$$\mu_N = (m_p \cdot \lambda_p^2 \cdot t_p^{-1}) \ (4 \ \pi)^{-1} \ q_p \cdot m_p^{-1} = (4 \ \pi)^{-1} \ q_p \cdot t_p^{-1} \cdot \lambda_p^2. \quad (106)$$

3. Discussion

Dividing the SI value of μ_B by that of μ_N gives:

$$\begin{aligned}
\beta &= \mu_B \ \mu_N^{-1} = (9.274 \times 10^{-24} \ \text{A·m}^2) \\
&\quad \times \ (5.050 \times 10^{-27} \ \text{A·m}^2)^{-1} \\
&= 1836.15\dots. \quad (107)
\end{aligned}$$

Dividing the SE value of μ_B by the SP value of μ_N gives:

$$\begin{aligned}
\beta &= \mu_B \ \mu_N^{-1} = [(4 \ \pi)^{-1} \ |q_e| \cdot t_e^{-1} \cdot \lambda_e^2] \ [(4 \ \pi)^{-1} \ |q_p| \cdot t_p^{-1} \cdot \lambda_p^2]^{-1} \\
&= [c \ (4 \ \pi)^{-1} \ |q_e| \cdot \lambda_e] \ [c \ (4 \ \pi)^{-1} \ |q_p| \cdot \lambda_p]^{-1} \\
&= \lambda_e \cdot \lambda_p^{-1} = (\lambda_e) \ (\beta^{-1} \ \lambda_e)^{-1} = 1836.15\dots. \quad (108)
\end{aligned}$$

The values of Eqs. 107 and 108 are the same.

P. Quantum-gravity phenomena

The following sections present *quantum* aspects of gravitational phenomena based upon the discoveries revealed in previous sections. When using SI units, the equations that pertain to gravitational phenomena contain Newton's gravitation constant G. When using SG units, G converts into a combination of mathematical factors of which the fine-structure constant is the most relevant one. This means that the magnitude of inertial force affects the magnitudes of these gravitational phenomena.

1. Acceleration of gravity using SI units

The historical SI equation for the magnitude of the acceleration of gravity on the surface of a homogeneous solid sphere [41] is:

$$g_m = G_m M_m r_m^{-2}, \tag{109}$$

where:

G_m uses SI values,

M_m = the mass of the sphere in kilograms, and

r_m = the radius of the sphere in meters.

2. Acceleration of gravity at the surface of Earth

We can calculate the SI value of the acceleration of gravity on the surface of Earth by knowing that the average density of Earth d_{Em} is 5.5×10^3 kg·m^{-3}. By the definition of the meter, we know that the circumference of Earth c_{Em} is 4.0×10^7 m and its radius r_{Em} is $4.0 \times 10^7 (2\pi)^{-1}$ m. Therefore, the SI volume of Earth is:

$$
\begin{aligned}
V_{Em} &= 4(3)^{-1} \pi \, r_{Em}^3 \\
&= 4(3)^{-1} \pi \, [(4.0 \times 10^7)(2\pi)^{-1} \, m]^3 \\
&= (4.0 \times 10^7 \, m)^3 (6\pi^2)^{-1} = 1.1 \times 10^{21} \, m^3.
\end{aligned} \tag{110}
$$

The SI mass of Earth M_{Em} is its average density times its volume:

$$
\begin{aligned}
M_{Em} = d_{Em} V_{Em} &= (5.5 \times 10^3 \, kg\cdot m^{-3})(1.1 \times 10^{21} \, m^3) \\
&= 6.1 \times 10^{24} \, kg.
\end{aligned} \tag{111}
$$

Using Eq. 111, the SI value of the magnitude of the acceleration of gravity at the surface of Earth is:

$$
\begin{aligned}
g_{Em} = G_m M_{Em} r_{Em}^{-2} &= (6.7 \times 10^{-11} \, kg^{-1}\cdot m^3\cdot s^{-2}) \\
\times (6.1 \times 10^{24} \, kg) &\, [4.0 \times 10^7 (2\pi)^{-1} \, m]^{-2} \\
&= 9.8 \, m\cdot s^{-2}.
\end{aligned} \tag{112}
$$

Notice that only two significant digits are used in the above calculations. This is because Earth is not a perfect sphere, and its density is not homogeneous, so the values cannot be very precise. Also, the magnitude of the acceleration of gravity at the surface of Earth varies over its surface.

3. Acceleration of gravity using SG units

The SG quantum acceleration equation is (see Table VII):

$$
\begin{aligned}
g_g &= G \ M_g \ r_g^{-2} = [2 \ \alpha \ (4 \ \pi)^{-1} \ \lambda_g^{3} \cdot m_g^{-1} \cdot t_g^{-2}] \\
&\quad \times \ (|M|_g \ m_g) \ (|r|_g^{-2} \ \lambda_g^{-2}) \\
&= 2 \ \alpha \ |M|_g \ (4 \ \pi \ |r|_g^{2})^{-1} \ \lambda_g \cdot t_g^{-2} \\
&= 2 \ \alpha \ |M|_g \ |S|_g^{-1} \ \lambda_g \cdot t_g^{-2},
\end{aligned}
\tag{113}
$$

where:
$|S|_g$ is the dimensionless whole-number SG area of the surface of the sphere, and
$|M|_g$ is the dimensionless whole-number SG mass of the sphere.

We can read this equation as: The SG value of the magnitude of the acceleration of gravity on the surface of a homogeneous sphere of matter is twice its mass divided by about 137 times the area of its surface. This means that the magnitude of the acceleration of gravity decreases as the magnitude of the fine-structure constant decreases. For example, in a dense part of the Universe where the magnitude of the fine-structure constant would be $1/274$ rather than our $1/137$, the magnitude of the acceleration of gravity would be 4.9 m·s^{-2} or half of Earth's 9.8 m·s^{-2}.

4. Escape velocity using SI units

The historical SI equation for the magnitude of the velocity of escape from the surface of a spherical mass [42] is:

$$
v_m = (2 \ G \ M_m \ r_m^{-1})^{0.5}.
\tag{114}
$$

5. Escape velocity of Earth

The equation for the velocity of escape from the surface of Earth is:

$$
\begin{aligned}
v_{Em} &= \{2 \ (6.7 \times 10^{-11} \ \text{kg}^{-1} \cdot \text{m}^3 \cdot \text{s}^{-2}) \ (6.1 \times 10^{24} \ \text{kg}) \\
&\quad \times [4.0 \times 10^7 \ (2 \ \pi)^{-1} \ \text{m}]^{-1}\}^{0.5} \\
&= 1.1 \times 10^4 \ \text{m·s}^{-1},
\end{aligned}
\tag{115}
$$

which is equal to approximately $25,000$ miles per hour.

6. Escape velocity using SG units

The equation for the SG value of the magnitude of the velocity of escape from the surface of a spherical mass is (see Table VII):

$$v_g = (2 \ G \ M_g \ r_g^{-1})^{0.5}$$
$$= \{2 \ [2 \ \alpha \ (4 \ \pi)^{-1} \ \lambda_g^3 \cdot m_g^{-1} \cdot t_g^{-2}] \ (|M|_g \ m_g) \ (|r|_g \ \lambda_g)^{-1})\}^{0.5}$$
$$= (\alpha \ \pi^{-1} \ |M|_g \ |r|_g^{-1})^{0.5} \ \lambda_g \cdot t_g^{-1}. \tag{116}$$

7. Schwarzschild radius using SI units

By definition, the Schwarzschild radius r_0 is the radius of a homogeneous, spherical of mass M that is compressed to the point that it becomes a Black Hole from which light cannot escape because the velocity of escape from the surface of the spherical mass would be greater than the speed of light, which is impossible. The historical SI equation for the speed-of-light escape velocity is derived from Eq. 114, where $(v_m = c)$:

$$c = (2 \ G_m \ M_m \ r_{0m}^{-1})^{0.5}. \tag{117}$$

The equation for the SI value of the Schwarzschild radius is:

$$r_{0m} = 2 \ G_m \ M_m \ c^{-2}. \tag{118}$$

8. Schwarzschild radius in SG units

The SG equation for the speed-of-light escape velocity is based upon Eq. 116, where $(v_g = c)$:

$$c = 1 \ \lambda_g \cdot t_g^{-1} = (\alpha \ \pi^{-1} \ |M|_g \ |r_0|_g^{-1})^{0.5} \ \lambda_g \cdot t_g^{-1}. \tag{119}$$

The equation for the SG value of the Schwarzschild radius is (see Table VII):

$$r_{0g} = 2 \ G_g \ M_g \ c^{-2} = 2 \ [2 \ \alpha \ (4 \ \pi)^{-1} \ \lambda_g^3 \cdot m_g^{-1} \cdot t_g^{-2}]$$
$$\times \ (|M|_g \ m_g) \ (\lambda_g \cdot t_g^{-1})^{-2} = \alpha \ \pi^{-1} \ |M|_g \ \lambda_g. \tag{120}$$

In essence, if a homogeneous, spherical entity's mass-to-radius ratio in SG units is less than $(\alpha \ \pi^{-1})$, the entity is a Black Hole. This equation shows that the magnitude of a Schwarzschild radius decreases as the magnitude of the fine-structure constant decreases.

9. Density of a Black Hole

The equation for the density d of a homogeneous sphere of radius r is mass M per volume V:

$$d = M \ V^{-1} = M \ [\tfrac{3}{4}^{-1} \ \pi \ r^3]^{-1}. \qquad (121)$$

The SG quantum equation for the density of a Black Hole with a Schwarzschild radius [44] is (see Table VII):

$$
\begin{aligned}
d_{0g} &= 3 \ M_g \ [4 \ \pi \ (r_{0g})^3]^{-1} \\
&= 3 \ (|M|_g \ m_g) \ [4 \ \pi \ (\alpha \ \pi^{-1} \ |M|_g \ \lambda_g)^3]^{-1} \\
&= \tfrac{3}{4} \ \pi^2 \ \alpha^{-3} \ |M|_g^{-2} \ m_g \cdot \lambda_g^{-3},
\end{aligned} \qquad (122)
$$

where the density is inversely proportional to the square of the mass. Therefore, we realize that the less massive the Black Hole, the more dense it must be. Notice that the powers of α and of λ_g are the same—at minus three.

10. Earth as a Black Hole

A Black Hole as massive as Earth possesses an SI Schwarzschild radius of:

$$
\begin{aligned}
r_{Em} &= 2 \ G \ M_{Em} \ c^{-2} = 2 \ (6.7 \times 10^{-11} \ kg^{-1} \cdot m^3 \cdot s^{-2}) \\
&\times (6.1 \times 10^{24} \ kg) \ (3.0 \times 10^8 \ m \cdot s^{-1})^{-2} \\
&= 9.0 \times 10^{-3} \ m.
\end{aligned} \qquad (123)
$$

Can you imagine—the mass of Earth contained within a ball of bubble gum? This seems to mean that the content of an atom in the vicinity of the Solar System is *really, really—mostly nothing*.

11. Black Hole as dense as water

The equation for the density of a Black Hole that is as dense as water is:

$$
\begin{aligned}
d &= 1.0 \times 10^3 \ kg \cdot m^{-3} \\
&= 3 \ \pi^2 \ \alpha^{-3} \ (4 \ |M|_g^2)^{-1} \ m_g \cdot \lambda_g^{-3}.
\end{aligned} \qquad (124)
$$

The SI unit of density $kg \cdot m^{-3}$ converts to the SG unit $m_g \cdot \lambda_g^{-3}$ by replacing the SI units with their values in SG units:

$$
\begin{aligned}
1 \ kg \cdot m^{-3} &= [(1.859 \times 10^{-9})^{-1} \ m_g] \ [(1.188 \times 10^{-33})^{-1} \ \lambda_g]^{-3} \\
&= 9.019 \times 10^{-91} \ m_g \cdot \lambda_g^{-3}.
\end{aligned} \qquad (125)
$$

Substituting the SG units for the SI units, gives:

$$d = (1.0 \times 10^3)(9.019 \times 10^{-91}) \; m_g \cdot \lambda_g^{-3}$$
$$= 3 \; \pi^2 \; \alpha^{-3} \; (4 \; |M|_g^2)^{-1} \; m_g \cdot \lambda_g^{-3}. \qquad (126)$$

Removing the units of measure to simplify, gives:

$$d = (9.1 \times 10^{-88}) = 3 \; \pi^2 \; \alpha^{-3} \; (4 \; |M|_g^2)^{-1}. \qquad (127)$$

Solving for the SG mass of that enormous Black Hole (from our perspective) gives:

$$|M|_g = [3 \; \pi^2 \; \alpha^{-3} \; (4)^{-1} \; (9.1 \times 10^{-88})^{-1}]^{0.5} = 1.4 \times 10^{47}$$
$$\text{or}$$
$$M_g = 1.4 \times 10^{47} \; m_g \qquad (128)$$

A Black Hole with a density of water would contain an SI mass of:

$$M_m = (1.4 \times 10^{47} \; m_g)(1.9 \times 10^{-9} \; kg \cdot m_g^{-1})$$
$$= 2.7 \times 10^{38} \; kg \qquad (129)$$

and is over a *hundred-million* times more massive than the Solar System. Its SI circumference is almost three times the length of the orbit of Earth about the Sun:

$$C_m = 4 \; \pi \; G \; M \; c^{-2} = 4 \; \pi \; (6.7 \times 10^{-11} \; kg^{-1} \cdot m^3 \cdot s^{-2})$$
$$\times (2.7 \times 10^{38} \; kg)(3.0 \times 10^8 \; m \cdot s^{-1})^{-2}$$
$$= 2.5 \times 10^{12} \; m. \qquad (130)$$

12. Temperature of a Black Hole

The SI equation for the temperature T of the event horizon of a Black Hole of mass M is:

$$T_m = \hbar \; c^3 \; (8 \; \pi \; k \; G \; M)^{-1}. \qquad (131)$$

The SG quantum equation for this temperature is (see Table VII):

$$T_g = [(2 \; \pi)^{-1} \; E_g \cdot t_g] \; (\lambda_g \cdot t_g^{-1})^3$$
$$\times \{[8 \; \pi] \; [E_g \cdot k_g^{-1}] \; [2 \; \alpha \; (4 \; \pi)^{-1} \; m_g^{-1} \cdot \lambda_g^3 \cdot t_g^{-2}] \; [|M|_g \; m_g]\}^{-1}$$
$$= (8 \; \pi \; \alpha \; |M|_g)^{-1} \; k_g. \qquad (132)$$

13. Black-Hole entropy using SI units

Beckenstein and Hawking's equation for the entropy of a Black Hole [40] using SI units, with a zero value for both charge ($Q=0$) and angular momentum ($J=0$), is:

$$S_m = A \, k \, c^3 \, (4 \, \hbar \, G)^{-1}, \tag{133}$$

where:
    A is the area of the event horizon of the Black Hole in meters squared.
    The constants k, c, $\hbar$, and G use SI values.

14. Black-Hole entropy using SG units

To convert Black-Hole entropy from using SI units to SG units, let:

$$A = |A|_g \, \lambda_g^2, \text{ where the value of } |A|_g \text{ is a dimensionless whole number,}$$
$$k = E_g \cdot k_g^{-1},$$
$$c^3 = \lambda_g^3 \cdot t_g^{-3}$$
$$\hbar = E_g \cdot t_g \, (2 \, \pi)^{-1},$$
$$G = 2 \, \alpha \, (4 \, \pi)^{-1} \, m_g^{-1} \cdot \lambda_g^3 \cdot t_g^{-2}.$$

Assembling the factors into the SG equation for the entropy of a Black Hole gives (see Table VII):

$$S_g = (|A|_g \, \lambda_g^2) \, (E_g \cdot k_g^{-1}) \, (\lambda_g^3 \cdot t_g^{-3})$$
$$\times \, \{ \, [4] \, [E_g \cdot t_g \, (2 \, \pi)^{-1}] \, [2 \, \alpha \, (4 \, \pi)^{-1} \, m_g^{-1} \cdot \lambda_g^3 \cdot t_g^{-2}] \, \}^{-1}$$
$$= \alpha^{-1} \, \pi^2 \, |A|_g \, m_g \cdot \lambda_g^2 \cdot t_g^{-2} \cdot k_g^{-1} = \alpha^{-1} \, \pi^2 \, |A|_g \, E_g \cdot k_g^{-1}$$
$$= \alpha^{-1} \, \pi^2 \, |A|_g \, S_g. \tag{134}$$

The SG *quantum* unit of entropy s_g is comprised of the SG quantum unit of energy E_g per SG quantum unit of temperature k_g which is the same as Boltzmann's constant k. When using a system of unit measures based upon the quantum attributes of the masstron, the multitude of constants in Hawking's equation for the entropy of a Black Hole vanish, and ($\alpha^{-1} \, \pi^2$) take their places. However, the π^2 factor looks suspicious; just π seems to be more reasonable. I possess this opinion because I do not believe that I have ever seen a formula for spacial phenomena that uses the *square* of π as a factor.

15. Discussion

All of the SI equations for the gravitational phenomena include G as a factor. Because the G constant includes α among its factors, α continues to occur in the SG quantum equations. However, G would not have contained α if gravitational forces had not been measured using a unit of inertial force (the newton), rather than a gravitational unit. The quantum-gravity phenomena show that the magnitude of the fine-structure constant affects the magnitudes of them all.

IV. OTHER CONSIDERATIONS

Here I present my ideas about two of my pet subjects, which I have thought about a lot over the decades: the structure of the Universe and the wave-particle paradox.

A. Dual-universe model

As the Universe first formed, apparently matter and antimatter conjugates were created at the same points in space and immediately gravitationally repulsed each other to pool into separate groups of matter and antimatter. Over time, these groups attracted more of their kind until they ignited into stars and antistars, which collected into galaxies and antigalaxies. These formed into clusters and anticlusters and, then, into superclusters and antisuperclusters, and on and on.

The Universe, it seems, is a dual universe, composed of equal quantities of matter and antimatter such that the sums of their respective attributes and anti-attributes equal zero. Matter possesses attributes or dimensions of mass, temperature, charge, length, and time. The anti-attributes of antimatter are antimass, antitemperature, anticharge, antilength, and antitime.

1. Galactic radiation

Possibly, antigalaxies emit the same kind of radiation as is emitted by galaxies. Electromagnetic radiation seems to be half matter and half antimatter—some form of interlocking occurrences of electrons and positrons at the speed of light (see Section III.L.).

As conjugate pairs in an energy mode of existence (see Section IV.A.4.), the mass of the electrons cancels the antimass of the positrons, so the energy beam's effective mass is zero. More-powerful protomagnetic radiation might be possible—consisting of half matter and half antimatter—in this case, interlocking occurrences of protons and antiprotons at the speed of light. Further, interlocking occurrences of masstrons and anti masstrons might produce gravitons, also, at the speed of light.

2. Force conjugates

Not only do attributes seem to be conjugates of anti-attributes, so do the various forces, of their respective antiforces. Although the force of gravitation and the force of antigravitation are attractive within their respective environments, reasonably, matter and antimatter *gravitationally* repel each other. The electromagnetic and anti-electromagnetic forces should follow similar but opposite rules. Electrons electromagnetically repel other electrons; positrons—other positrons, but electrons attract positrons. In sum, within each of the two environments, particles that are alike, gravitationally attract and electromagnetically repel, while, when spanning the two, the opposite appears to be true —conjugate particles gravitationally repel and electromagnetically attract.

3. Age and entropy conjugates

In this dual model of the Universe, *symmetry* requires that the age of the matter part of the dual universe is equal and opposite to the anti-age of the antimatter part. Added together, the age of the entire dual universe is zero and appears to be ageless. Its total entropy is zero, as well. Although entropy and anti-entropy both increase over time, anti-entropy increases over antitime, so, as far as the matter world is concerned, it decreases and cancels out entropy for the dual universe.

4. Modes of existence

Apparently, matter and antimatter can coexist in one of three modes of existence—discrete, energy, or virtual—depending upon the distance between matter and antimatter elementary particles.

- In the discrete mode, a particle and an antiparticle are far enough apart to remain as discrete particles.
- In the energy mode, a particle and an antiparticle are close enough to form an energy beam, but they do not exactly coincide in space. A distance—the wavelength of the energy beam—separates them.
- In the virtual mode, the wavelength is zero. The matter and antimatter particles coincide spatially and virtually disappear, leaving nothing behind, not even energy. The sum of their attributes and anti-attributes is zero—mass cancels antimass, temperature—antitemperature, time—antitime, space—antispace, and so on. Nothing remains, not even empty space—not even a void.

B. Wave-particle duality

The content of this section refutes the theory about wave-particle duality and proposes an alternative pure-corpuscular theory, which seems to be more reasonable and plausible. Everything is composed of particles, yet they may congregate in the crest of waves and avoid their troughs, or they may disperse themselves in a linear pattern of fields—but, still, they are only *particles*.

1. Background

Two centuries ago, long before the discovery of quantum phenomena, Augustin Fresnel and Thomas Young[45] projected beams of light through slits in a thin light barrier onto a screen beyond, where parallel lines appeared. Thirty-five years later, George Airy,[46] while studying the characteristics of telescopes, encountered concentric rings of light. Fresnel and Young interpreted the cause of the parallel lines to be wave interference.[47] Airy came to the same conclusion for his concentric rings of light. None of them could have come to any other conclusion because the discontinuous quantum nature of the attributes of elementary particles was unknown during their lifetimes. Thus was born the theory of the particle-wave duality of matter,[48] which has been maintained and reënforced throughout the Nineteenth and Twentieth centuries and now into the Twenty-first by fitting "proof" after "proof" to these preconceived notions to "verify" of what the observed physical phenomena seemed to represent.

2. Quantum diffraction angles

Based upon discoveries explained in Section III., especially the factorization of Planck's constant into quantum attributes of the electron (see Section III.A.), perhaps quantum diffraction angles cause these phenomena, not wave interference. This is so because, on the quantum level, only elementary particles appear to exist, not waves or fields, which seem to be only "invented" phenomena to explain the reality of classical physics. Waves appear to show a frontal movement of masses of *particles* in their alternating occurrences of denseness and sparseness. Fields, in turn, show the direction and concentration of a constant rectilinear flow of *particles*. So, particles exist, but waves and fields were "invented" just to describe the concentration or ebb and flow of an assembly of *particles*.

The quantum effect of diffraction is observed when a high-velocity elementary particle diffracts off an edge of a slit or a pinhole in a thin light barrier to hit a screen beyond.[49] This edge, in any experiment, is made as fine and regular as is technologically possible, yet, because of physical limitations in accuracy tolerances on the quantum scale, it remains rough in comparison to the particle diffracting from it. If each particle could be made to hit the edge in the exact same way at the exact same spot every time, then the diffraction angle would always be the same. The experiment can control neither the edge roughness nor the aim of the particle this closely; therefore, each particle's trajectory changes by a random diffraction angle.

However, each particle does *not* appear to diffract at a completely random angle but, randomly, within a *discontinuous quantization of the angles*. In essence, an elementary particle's diffraction from a slit or pinhole deflects its trajectory by a *whole-number multiple of a quantum angle*, which can only represent a combination of that particle's quantum attributes of length and time. Once diffracted, if the particle is disturbed—such as in an attempt to measure it—its trajectory again changes by one or more of that particle's quantum angle, but, because this disturbance would usually occur at a random location, the angle of the particle's new trajectory would not be a whole-number multiple of its previous angle. This means that the point of destination would be random, and parallel lines or concentric rings would not appear on the destination screen.

When a double-slit apparatus possesses a particular physical configuration in harmony with the wavelength of the elementary particle, those particles that diffract from one slit could focus on the destination screen in phase with those diffracting from the second slit. Apparently, this would reinforce the parallel-line phenomenon such that, when one slit is covered, this reinforcement would cease and the parallel lines would appear to vanish.

In addition to diffraction, perhaps the same quantum-angle phenomenon affects reflection and refraction, yet, because these phenomena deal with beams of particles with a broad front (plane "waves"), the effects of parallel-beam transmission may hide them. Other reflection and refraction effects that, possibly, can be attributed to the discontinuous, quantum nature of matter are: iridescence, diffraction gratings, partial reflection, and holograms.[50]

3. Wave-particle-theory legacy

Augustin Fresnel, Thomas Young, and George Airy explained the cause of the phenomena that they observed as best they could in a classical (Newtonian) world a century before awareness of quantum discontinuity occurred. Until the time of Planck and Einstein, wave interference was the only possible explanation. Yet, the succession of wave-interference "proofs" so imprinted themselves on the minds of scientists, one generation after the other, that the possibility of searching for another (the correct) solution did not seem to occur to them.

Even the founders of quantum theory, each in his own way, tried to further justify wave-particle duality. Erwin Schrödinger's wave mechanics [51] was successful in bolstering this concept. Yet, its success further discouraged new searches for the real solution to the wave-particle paradox. Instead, it helped to spawn the Copenhagen interpretation,[52] the EPR paradox,[53] Bohm's hidden variables,[54] the Bell theorem,[55] Aspect's experiment,[56] Everett's parallel worlds, and so on. These events suggested the existence of superluminal signaling,[57] nonlocality of cause and effect,[58] parallel universes,[59] and the creation of whole galaxies [60] by just thinking about them.

V. CONCLUSION

Certainly, in the future, these various Twentieth-century ideas about quantum theories will be classified with those of centuries past—such as Earth being the center of the Universe. Perhaps, Schrödinger's wave equations do the job, but, to this day, to a certain extent, so does the astrolabe.

Nick Herbert describes the quantum-theory situation perfectly in an allegory at the beginning of Chapter 9 of his book, *Quantum Reality*.[61] He puts it so well, and it is so relevant to this essay, that just a reference to it is not sufficient. Quoting from Herbert's book:

"Quantum theory resembles an elaborate tower whose middle stories are complete and occupied. Most of the workmen are crowded together on top, making plans and pouring forms for the next stories. Meanwhile the building's foundation consists of the same temporary scaffolding that was rigged up to get the project started. Although he must pass through them to get to the rest of the city, the average physicist shuns these lower floors with a kind of superstitious dread. New York University professor Daniel Greenberger, speaking at a recent *festschrift*, speculated about why most physicists avoid the quantum reality question:

" 'This sudden success on a grand scale, after a generation of desperate striving by great minds, lends a heroic, even mystic, quality to the history of [quantum theory]. But, inevitably, it has also led to a sensitivity on the part of physicists, a kind of defensiveness, ultimately arising from the fear that the whole delicate structure, so painstakingly put together, might crumble if touched. This has tended to produce a "Let's leave well enough alone" attitude, which I believe contributes to the great reluctance most physicists have to tinker with, or even critically examine, the foundations of quantum theory. However, fifty years have gone by and the structure appears stronger than ever.'

"Physicists' reality crisis consists of the fact that nobody can agree on what's holding the building up. Different people looking at the same theory come up with profoundly different models of reality, all of them outlandish compared to the ordinary experience which constitutes both daily life and the quantum facts. Physicists differ over which parts of this theory they will take seriously and which parts they will ignore as empty formalism having no counterpart in the real world. Which different picture of quantum reality you end up with depends on what parts of quantum theory you take seriously. In this chapter and the next I examine how the eight major quantum realities arise from the selective emphasis of certain features of quantum theory and the neglect of others." [end of the quote from Herbert's book]

Perhaps, the contents of this essay will be a first step in prompting scientists to descend Herbert's "elaborate tower" for an analysis of its foundation's integrity and for its eventual restructuring in harmony with the dis-coveries presented in this essay. Hopefully, it will enable scientists to create less-arcane and more-meaningful quantum equations, free of confusing and irrelevant "constants." This will lead them to discovering more truths about the physical environment on both the quantum and cosmic scales, such as this essay's revelation as to the real significance of the magnitude of the fine-structure constant—a number *a little bit bigger than 137.*

VI. REFERENCES

[1] Z. Yan and G. Drake, Phys. Rev. Lett. **74** (24), 4791 (1995). M. Inguscio, G. Guisfredi, F. Pavone, and M. Prevedelli, Physica Scripta **T70**, 42 (1997).

[2] J. Thomson, Phil. Mag. **44**, 295 (1897). Ibid., "Recollections and Reflections" (G. Bell and Sons, London 1936), p. 341. P. Zeeman, Nature **55**, 347 (1897).

[3] S. Weinberg, "The First Three Minutes, updated edition" (Basic Books, Inc., New York 1988), p. 82, 156. H. Pagels, "Perfect Symmetry: The Search for the Beginning of Time" (Simon and Schuster, New York 1985), p. 240, 252. J. Trefil, "Science Digest," p. 55 (June 1984).

[4] R. Millikan, Phys. Rev. **8**, 595, (1916).

[5] A. Compton, Phys. Rev. **21**, 207, (1923). Ibid., 715. Ibid., **23**, 763, (1924). D. Bohm, "Quantum Theory" (Prentice-Hall, Englewood Cliffs, NJ 1951) p. 33.

[6] P. Davies, "The Forces of Nature, 2nd ed." (Cambridge University Press, New York, 1986), p. 88.

[7] G. Cohen-Tannoudji, "Universal Constants in Physics" (McGraw-Hill, New York 1993) [translated from "Les Constantes universelles" (Hachette, Paris 1991)], p. 54.

[8] W. Heisenberg, "Principles of Quantum Physics [translated by C. Eckart and F. Hoyt] (Dover Publications, New York 1930). D. Bohm, ibid.,[5] p. 99. R. Feynman, R. Leighton, and M. Sands, "The Feynman Lectures on Physics: Quantum Mechanics (vol. III)," third printing (July 1966), (Addison-Wesley, Reading, MA 1965), p. 1-11.

[9] D. Halliday, R. Resnick, and K. Krane, "Physics, 4th Edition" (John Wiley & Sons, Inc., New York, 1992), vol. 1, p. 85.

[10] R. Feynman, "The Character of Physical Law," 23rd printing (1998), (The MIT Press, Cambridge, MA), p. 31.

[11] Ibid.,[9] vol. 2 extended, p. 597.

[12] J. Daintith, *et al.*, "Penguin Dictionary of Physics" (Edited by V. Illingworth), (Penguin Books Ltd., London 1991), p. 343.

[13] D. Halliday, R. Resnick, and K. Krane, ibid.,[9] vol. 2 extended, p. 767.

[14] J. Daintith, *et al.*, ibid.,[12] p. 343.

[15] D. Halliday, R. Resnick, and K. Krane, ibid.,[9] vol. 2 extended, p. 879.

[16] Ibid.,[9] vol. 1, p. 345.

[17] J. Daintith, *et al.*, ibid.,[12] p. 201.

[18] Ibid.,[12] p. 354.

[19] A. Landé, Phys. Rev. **59**, 434 (1941). M. Luban and H. Chew, Phys. Rev. D **31**, 2643 (1985).

[20] N. Bohr, D. Kgl. Danske Vidensk. Selsk. Skrifter, Naturvidensk. og Mathem. **8**, IV.1, 1-3 (1918).

[21] D. Bohm, ibid.,[5] p. 44. D. Halliday, R. Resnick, and K. Krane, ibid.,[9] vol. 2 extended, p. 1073.

[22] D. Ingram, "Radiation and Quantum Physics" (Clarendon Press, Oxford 1973) p. 24.

[23] Ibid.,[22] p. 41.

[24] J. Daintith, et al., ibid.,[12] p. 451. N. Calder, "Einstein's Universe" (Penguin Books, New York 1983), p. 173.

[25] D. Halliday, R. Resnick, and K. Krane, ibid.,[9] vol. 2 extended, p. 1145. N. Calder, ibid.,[24] p. 32.

[26] A. Einstein, Phys. Zs. **18**, 121 (1916).

[27] G. Ohm, "Die galvanische Kette, mathematisch bearbeitet" (Köln 1827).

[28] D. Halliday, R. Resnick, and K. Krane, ibid.,[9] vol. 2 extended, p. 654.

[29] Ibid.,[9] vol. 2 extended, p. 698.

[30] D. Dominguez, K. Von Klitzing, and K. Ploog, Metrologia **26**, 197 (1989). K. Von Klitzing, Physica B **204**, 111 (1995).

[31] B. Halperin, "The Quantized Hall Effect" (Scientific American, April 1986), p.52.

[32] B. Deaver and W. Fairbank, Phys. Rev. Lett. **7**, 43 (1961). R Döll and M. Nabauer, ibid. **7**, 51 (1961).

[33] L. Boltzmann, "Vorlesungen über Gastheorie" (Barth, Leipzig 1896). R. Christy and A. Pytte, "the structure of matter: an introduction to modern physics" (W. A. Benjamin, Inc., New York 1965), p. 116.

[34] R. Penrose, "Big Bangs, Black Holes and 'Time's Arrow'" in "The Nature of Time" (Edited by R. Flood and M. Lockwood) (Basil Blackwell, New York 1986), p. 52. H. Pagels, "Perfect Symmetry: The Search for the Beginning of Time" (Simon and Schuster, New York 1985), p. 72.

[35] A. Avogadro, Journal de Physique **73**, 58 (1811).

[36] J. Daintith, et al., ibid.,[12] p. 25.

[37] R. Boyle, "New Experiments Physio-Mechanicall [sic], Touching the Spring of the Air and its Effects" (London, 1661). J. Gay-Lussac, Mémoires de la Société d'Arcueil **2**, 207 (1809).

[38] J. Daintith, et al., ibid.,[12] p. 284.

[39] Ibid.[12].

[40] J. Beckenstein, Phys. Rev. D **7**, 2333 (1973). Ibid., **9**, 3292 (1974). S. Hawking, Comm. in Math. and Phys. **43**, 199 (1975).

[41] D. Halliday, R. Resnick, and K. Krane, ibid.,[9] vol. 1, p. 348.

[42] Ibid.,[9] vol. 1, p 354.

[43] K. Schwarzschild, Vierteljahrschr. der Astrom. Ges. **35**, 337 (1900). P. Peebles, "Principles of Physical Cosmology" (Princeton University Press, Princeton, NJ 1993), p. 289.

[44] S. Hawking, "A Brief History of Time: from the Big Bang to Black Holes" (Bantum Books, New York 1988), p. 81.

[45] P. Davies, "Time Asymmetry and Quantum Mechanics" in "The Nature of Time," ibid.,[6] p. 103. T. Hey and P. Walters, "The quantum universe" (Cambridge University Press, Cambridge 1987), p. 3, 6.

[46] N. Herbert, "Quantum Reality: beyond the new physics" (Anchor Press/Doubleday, Garden City, NY 1985), p. 63.

[47] R. Feynman, R. Leighton, and M. Sands, ibid.,[8] p. 1.

[48] D. Bohm, ibid.,[5] p. 116.

[49] T. Hey and P. Walters, ibid.,[45] p. 6, 12. J. Polkinghorne, "The Quantum World" (Princeton University Press, Princeton, NJ 1989), p. 34.

[50] R. Feynman, "QED: the strange theory of light and matter" (Princeton University Press, Princeton, NJ, 1985), p. 34, 45, 100.

[51] E. Schrödinger, Phys. Rev. **28**, 1049 (1926).

[52] N. Bohr, Nature **121**, 580 (1928).

[53] A. Einstein, B. Podolsky, and N. Rosen, Phys. Rev. **47**, 777 (1935). N. Bohr, Phys. Rev., **48**, 696, (1935). W. Furry, Phys. Rev., **49**, 393, 476 (1936).

[54] D. Bohm, Phys. Rev. **85**, 155 (1952). J. Bell, Reviews of Modern Physics **38**, 447 (1966).

[55] J. Bell, Physics, **1**, 195 (1964).

[56] A. Aspect, J. Dalibard, and G. Roger, Phys. Rev. Lett., **49**, 91, 1804 (1982).

[57] S. Weinberg, Phys. Rev. Lett. **62**, 485 (1989).

[58] J. Polchinski, Phys. Rev. Lett. **66**, 397 (1991). F. Nogueira, M. Caldeira, and J. Domingos, Physica Scripta, **53** (1), 18 (1996).

[59] F. Wolf, "Parallel Universes: The Search for Other Worlds" (Simon and Schuster, New York 1988), p. 25.

[60] J. Cramer, Phys. Rev. D **22**, 362 (1980).

[61] N. Herbert, ibid.,[46] p. 158.

VII. TABLES

Table I. Electron quantum-attribute values—without the charge quantum

Name of the quantum attribute of the electron	SE code	SE value	Quantum attribute factor combination	SI value to 4 significant digits
mass	m_e	1	m_e	9.109×10^{-31} kg
threshold temperature	k_e	1	k_e	5.929×10^9 K
charge	q_e	1	q_e	1.602×10^{-19} C
Compton wavelength	λ_e	1	λ_e	2.426×10^{-12} m
virtual-electron lifetime	t_e	1	t_e	8.093×10^{-21} s
area	A, S	1	λ_e^2	5.886×10^{-24} m^2
volume	V	1	λ_e^3	1.428×10^{-35} m^3
cyclic frequency	ν	1	t_e^{-1}	1.235×10^{20} Hz
angular velocity	ω	1	t_e^{-1}	1.235×10^{20} radian·s^{-1}
angular acceleration	α	1	t_e^{-2}	1.526×10^{40} radian·s^{-2}
linear velocity	v	1	$\lambda_e \cdot t_e^{-1}$	2.997×10^9 m·s^{-1}
linear acceleration	a	1	$\lambda_e \cdot t_e^{-2}$	3.704×10^{-28} m·s^{-2}
mass density	ρ	1	$m_e \cdot \lambda_e^{-3}$	6.377×10^4 kg·m^{-3}
angular inertia	I	1	$m_e \cdot \lambda_e^2$	5.362×10^{-54} kg·m^2
pressure	p	1	$m_e \cdot \lambda_e^{-1} \cdot t_e^{-2}$	5.731×10^{21} Pa
inertial force (m_e **a**)	F	1	$m_e \cdot \lambda_e \cdot t_e^{-2}$	3.374×10^{-2} N

Table I (concluded). Electron quantum-attribute values—without the charge quantum

Name of the quantum attribute of the electron	SE code	SE value	Quantum attribute factor combination	SI value to 4 significant digits
torque (angular work)	τ	1	$m_e \cdot \lambda_e^2 \cdot t_e^{-2}$	8.187×10^{-14} N·m
energy, heat, linear work	$E, Q,$	1	$m_e \cdot \lambda_e^2 \cdot t_e^{-2}$	8.187×10^{-14} J
power	P	1	$m_e \cdot \lambda_e^2 \cdot t_e^{-3}$	1.011×10^7 W
power density	S	1	$m_e \cdot t_e^{-3}$	1.718×10^7 W·m^{-2}
linear momentum ($m_e\, v$)	p	1	$m_e \cdot \lambda_e \cdot t_e^{-1}$	2.730×10^{-22} kg·m·s^{-1}
angular momentum ($I\, \omega$)	L	1	$m_e \cdot \lambda_e^2 \cdot t_e^{-1}$	6.626×10^{-34} kg·m^2·s^{-1}

Table II. Electron quantum-attribute values—with the charge quantum

Name of the quantum attribute of the electron	SE code	SE value	Quantum attribute factor combination	SI value to 4 significant digits
electric current	I	1	$q_c \cdot t_c^{-1} = V R^{-1}$	1.979×10^{-1} A (C·s^{-1})
electric-current density	j	1	$q_c \cdot \lambda_c^{-2} \cdot t_c^{-1} = I \lambda_c^{-2}$	3.362×10^{24} A·m^{-2}
electric displacement	D	1	$q_c \cdot \lambda_c^{-2} = j\, t_c$	2.721×10^4 C·m^{-2}
electric potential	V	1	$m_c \cdot q_c^{-1} \lambda_c^2\, t_c^{-2} = I R$	5.109×10^5 V
electric resistance	R	1	$m_c \cdot q_c^{-2} \lambda_c^2 \cdot t_c^{-1} = G^{-1}$	2.581×10^4 Ω
electric conductance	G	1	$m_c^{-1} \cdot q_c^2 \cdot \lambda_c^{-2} \cdot t_c = R^{-1}$	3.874×10^{-5} S
electric resistivity ($\lambda_c R$)	ρ	1	$m_c \cdot q_c^{-2} \cdot \lambda_c^3 \cdot t_c^{-1} = \sigma^{-1}$	6.262×10^{-8} Ω·m
electric conductivity ($\lambda_c^{-1} G$)	σ	1	$m_c^{-1} \cdot q_c^2 \cdot \lambda_c^{-3} \cdot t_c = \rho^{-1}$	1.596×10^{-7} S·m^{-1}
electric capacitance	C	1	$m_c^{-1} \cdot q_c^2 \cdot \lambda_c^{-2} \cdot t_c^2 = t_c^2 L^{-1}$	3.135×10^{-25} F
magnetic inductance	L	1	$m_c \cdot q_c^{-2} \cdot \lambda_c^2 = t_c^2 C^{-1}$	2.089×10^{-16} H
electric flux	Φ_E	1	$m_c \cdot q_c^{-1} \cdot \lambda_c^3 \cdot t_c^{-2} = \boldsymbol{\nu}\, \Phi_B$	1.239×10^{-6} V·m
magnetic flux	Φ_B	1	$m_c \cdot q_c^{-1} \cdot \lambda_c^2 \cdot t_c^{-1} = \boldsymbol{\nu}^{-1} \Phi_E$	4.135×10^{-15} V·s
electric flux density	E	1	$m_c \cdot q_c^{-1} \cdot \lambda_c \cdot t_c^{-2} = \Phi_E\, \lambda_c^{-2}$	2.106×10^{17} V·m^{-1}
magnetic flux density	B	1	$m_c \cdot q_c^{-1} \cdot t_c^{-1} = \Phi_B\, \lambda_c^{-2}$	7.025×10^8 T
electric dipole moment	p	1	$q_c \cdot \lambda_c = \boldsymbol{\nu}^{-1} \mu$	3.887×10^{-31} C·m
magnetic dipole moment	μ	1	$q_c \cdot \lambda_c^2 \cdot t_c^{-1} = \boldsymbol{\nu}\, p$	1.165×10^{-22} A·m^2

Table II (concluded). Electron quantum-attribute values—with the charge quantum

Name of the quantum attribute of the electron	SE code	SE value	Quantum attribute factor combination	SI value to 4 significant digits
electric polarization	P	1	$q_e \cdot \lambda_e^{-2} = \nu^{-1} M$	2.721×10^{4} C·m^{-2}
magnetic magnetization	M	1	$q_e \cdot \lambda_e^{-1} \cdot t_e^{-1} = \nu P$	8.159×10^{12} A·m^{-1}
electric permittivity	$\in$	1	$m_e^{-1} \cdot q_e^{2} \cdot \lambda_e^{-3} \cdot t_e^{2} = (\nu^{2} \mu)^{-1}$	1.292×10^{-13} F·m^{-1}
magnetic permeability	μ	1	$m_e \cdot q_e^{-2} \cdot \lambda_e = (\nu^{2} \in)^{-1}$	8.610×10^{-5} H·m^{-1}
entropy	S	1	$m_e \cdot k_e^{-1} \cdot \lambda_e^{2} \cdot t_e^{-2} = E\, k_e^{-1}$	1.380×10^{-23} J·K^{-1}

Table III. Proton quantum-attribute values—comparing SP unity values to SE and SI

Name of the quantum attribute of the proton	SP code	SP value	SE value of the same magnitude	SI value of the same magnitude
proton-to-electron attribute conversion factor	β	1836.2	1836.15	1836.15
proton mass	m_p	1	$\beta \, m_e$	1.672×10^{-27} kg
proton threshold temperature	k_p	1	$\beta \, k_e$	1.088×10^{13} K
proton charge	q_p	1	$-q_e$	-1.602×10^{-19} C
proton Compton wavelength	λ_p	1	$\lambda_e \, \beta^{-1}$	1.321×10^{-15} m
virtual-proton lifetime	t_p	1	$t_e \, \beta^{-1}$	4.407×10^{-24} s

Table IV. Masstron quantum-attribute values—comparing SG values to SE and SI values

Name of the quantum attribute of the masstron	SG code	SG value	SE value of the same magnitude	SI value of the same magnitude
masstron-to-electron attribute conversion factor	γ	2.041×10^{21}	2.041×10^{21}	2.041×10^{21}
masstron mass	m_g	1	$\gamma\, m_e$	1.859×10^{-9} kg
masstron energy	E_g	1	$\gamma\, m_e \cdot \lambda_e^2 \cdot t_e^{-2}$	1.671×10^{8} J
masstron threshold temperature	k_g	1	$\gamma\, k_e$	1.210×10^{31} K
masstron Compton wavelength	λ_g	1	$\gamma^{-1} \lambda_e$	1.188×10^{-33} m
virtual-masstron lifetime	t_g	1	$\gamma^{-1} t_e$	3.965×10^{-42} s

Table V. Planck values compared to masstron quantum-attribute values

Dimension of the Planck Value	Planck code	Historical formula	Formula using quantum masstronic-attribute values	SI value of the Planck value
mass	M_{Pl}	$(\hbar\, c\, G^{-1})^{1/2}$	$\alpha^{-1/2}\, m_g$	2.176×10^{-8} kg
energy	E_{Pl}	$(\hbar\, c^5\, G^{-1})^{1/2}$	$\alpha^{-1/2}\, m_g \cdot \lambda_g^{2} \cdot t_g^{-2}$	1.956×10^{9} J
temperature	K_{Pl}	$(\hbar\, c^5\, G^{-1}\, k^{-2})^{1/2}$	$\alpha^{-1/2}\, k_g$	1.416×10^{32} K
length	L_{Pl}	$(\hbar\, c^{-3}\, G)^{1/2}$	$\alpha^{1/2}\,(2\pi)^{-1}\, \lambda_g$	1.616×10^{-35} m
lifetime	T_{Pl}	$(\hbar\, c^{-5}\, G)^{1/2}$	$\alpha^{1/2}\,(2\pi)^{-1}\, t_g$	5.390×10^{-44} s

Table VI. Electron-based fundamental universal physical constants of nature

Constant name	SE code	SE value	Various formulas	SI value to 4 significant digits
fine-structure constant	α	$(137...)^{-1}$	$q_e^2 (2\, \epsilon_0\, h\, c)^{-1}$	$(137.036)^{-1}$
speed of light	c	1	$\lambda_e \cdot t_e^{-1}$	$299\ 792\ 458$ m·s^{-1}
Planck constant	h	1	$m_e \cdot \lambda_e^2 \cdot t_e^{-1} = m_e \cdot \lambda_e\, c$	6.626×10^{-12} J·s
electron rest-mass energy	E_e	1	$m_e \cdot \lambda_e^2 \cdot t_e^{-2} = m_e\, c^2$	8.187×10^{-14} J
atomic mass unit (amu)	m_u	1822.89	$1822.89\ m_e$	1.660×10^{-27} kg
Boltzmann constant	k	1	$m_e \cdot k_e^{-1} \cdot \lambda_e^2 \cdot t_e^{-2} = E_e \cdot k_e^{-1}$	1.380×10^{-23} J·K^{-1}
universal gas constant	R	1	$m_e \cdot k_e^{-1} \cdot \lambda_e^2 \cdot t_e^{-2} = E_e \cdot k_e^{-1}$	8.314×10^3 J·K^{-1}·atoms·kmole^{-1}
Avogadro's number	N_A	1	(dimensionless)	6.022×10^{26} atoms·kmole^{-1}
Wien second radiation	c_2	1	$k_e \cdot \lambda_e$	1.438×10^{-2} m·K
electric current	I_e	1	$q_e \cdot t_e^{-1} = V_e\, R_K^{-1}$	1.979×10^{-1} m·A
Hall potential	V_e	1	$m_e \cdot q_e^{-1} \cdot \lambda_e^2 \cdot t_e^{-2}$	5.109×10^5 V
Hall resistance (von Klitzing)	R_K	1	$m_e \cdot q_e^{-2} \cdot \lambda_e^2 \cdot t_e^{-1} = G_e^{-1}$	2.581×10^4 Ω
Hall conductance	G_e	1	$m_e^{-1} \cdot q_e^2 \cdot \lambda_e^{-2} \cdot t_e = R_K^{-1}$	3.874×10^{-5} S
Hall resistivity ($\lambda_e\, R_K$)	ρ_e	1	$m_e \cdot q_e^{-2} \cdot \lambda_e^3 \cdot t_e^{-1} = \sigma_e^{-1}$	6.262×10^{-8} Ω·m
Hall conductivity ($\lambda_e^{-1}\, G_e$)	σ_e	1	$m_e^{-1} \cdot q_e^2 \cdot \lambda_e^{-3} \cdot t_e = \rho_e^{-1}$	1.596×10^{-7} S·m^{-1}
Bohr magneton	μ_B	$(4\pi)^{-1}$	$(4\pi)^{-1} q_e \cdot \lambda_e^2 \cdot t_e^{-1}$	9.274×10^{-12} J·T^{-1}

Table VI (concluded). Electron-based fundamental universal physical constants of nature

Constant name	SE code	SE value	Various formulas	SI value to 4 significant digits
permittivity constant	ϵ_0	$(2\,\alpha)^{-1}$	$(2\,\alpha)^{-1} \cdot f_{ic}^{-1} \cdot q_e^2 \cdot \lambda_e^{-2}$	8.854×10^{-12} F·m⁻¹ or N⁻¹·C²·m⁻²
permeability constant	μ_0	$2\,\alpha$	$2\,\alpha\, f_{ic}\, q_e^{-2}\, t_e^2$	1.256×10^{-6} H·m⁻¹ or N·C⁻²·s²
Wien first radiation	c_1	$2\,\pi$	$2\,\pi\; m_e \cdot \lambda_e^4 \cdot t_e^{-3}$	3.741×10^{-16} W·m²
Stefan-Boltzmann constant	σ	$2\,\pi^5\,15^{-1}$	$2\,\pi^5\,15^{-1}\; m_e \cdot k_e^{-4} \cdot t_e^{-3}$	5.672×10^{-8} W·m⁻²·K⁻⁴
Rydberg constant	R_∞	$\tfrac{1}{2}\,\alpha^2$	$\tfrac{1}{2}\,\alpha^2\,\lambda_e^{-1}$	1.097×10^7 m⁻¹
Rydberg frequency	ν_R	$\tfrac{1}{2}\,\alpha^2$	$\tfrac{1}{2}\,\alpha^2\,t_e^{-1}$	3.288×10^{15} s⁻¹
Rydberg energy	E_R	$\alpha^2\,2^{-1}$	$\alpha^2\,2^{-1}\; m_e \cdot \lambda_e^2 \cdot t_e^{-2} = 2^{-1}E_H$	2.179×10^{-18} J
Hartree energy	E_H	α^2	$\alpha^2\; m_e \cdot \lambda_e^2 \cdot t_e^{-2} = 2\,E_R$	4.358×10^{-18} J
Zeeman splitting constant	Z_s	$(4\,\pi)^{-1}$	$(4\,\pi)^{-1}\; m_e^{-1} \cdot q_e \cdot \lambda_e^{-1} \cdot t_e$	46.68 m·Wb⁻¹
classical electron radius	r_e	$\alpha\,(2\,\pi)^{-1}$	$\alpha\,(2\,\pi)^{-1}\,\lambda_e$	2.817×10^{-15} m
Wien displacement law	b_e	0.2014	$0.2014\, k_e \cdot \lambda_e$	2.897×10^{-3} K·m

Table VII. Masstron-based fundamental universal physical constants of nature

Constant name	SG Code	SG value	Various formulas	SI value to 4 significant digits
speed of light	c	1	$\lambda_g \cdot t_g^{-1}$	$299\ 792\ 458\ \text{m·s}^{-1}$
Planck constant	h	1	$m \cdot \lambda_g^2 \cdot t_g^{-1} = m_g \cdot \lambda_g\, c = E_g \cdot t_g$	$6.626 \times 10^{-12}\ \text{J·s}$
masstron rest-mass energy	E_g	1	$m_g \cdot \lambda_g^2 \cdot t_g^{-2} = m_g\, c^2 = E_g$	$1.671 \times 10^{8}\ \text{J}$
Boltzmann constant (entropy)	k	1	$m_g \cdot k_g^{-1} \lambda_g^2 \cdot t_g^{-2} = E_g \cdot k_g^{-1}$	$1.380 \times 10^{-23}\ \text{J·K}^{-1}$
Newton gravitation constant	G	$2\,\alpha\,(4\pi)^{-1}$	$2\,\alpha\,(4\pi)^{-1} f_{ig} \cdot m_g^{-2} \cdot \lambda_g^2$	$6.674 \times 10^{-11}\ \text{kg}^{-1}\text{·m}^3\text{·s}^{-2}$
black-hole entropy (area A)	s_g	$\alpha^{-1}\pi^2 \,\lvert A \rvert_g$	$\alpha^{-1}\pi^2 \,\lvert A \rvert_g\, m_g \cdot k_g^{-1}\,\lambda_g^2 \cdot t_g^{-2}$	$\lvert A \rvert_m \times 1.867 \times 10^{-20}\ \text{J·K}^{-1}$
gravity acceleration (mass M, sphere surface area S)	g_g	$2\,\alpha\,\lvert M \rvert_g\,\lvert S \rvert_g^{-1}$	$2\,\alpha\,\lvert M \rvert_g\,\lvert S \rvert_g^{-1}\,\lambda_g \cdot t_g^{-2}$	$\lvert M \rvert_m\,\lvert S \rvert_m^{-1} \times 6.674 \times 10^{-11}\ \text{m·s}^{-2}$
escape velocity (mass M, sphere radius r)	v_g	$(\alpha\,\pi^{-1}\,\lvert M \rvert_g\,\lvert r \rvert_g^{-1})^{\frac{1}{2}}$	$(\alpha\,\pi^{-1}\,\lvert M \rvert_g\,\lvert r \rvert_g^{-1})^{\frac{1}{2}}\,\lambda_g \cdot t_g^{-1}$	$(\lvert M \rvert_m\,\lvert r \rvert_m^{-1})^{\frac{1}{2}} \times 1.155 \times 10^{-5}\ \text{m·s}^{-1}$
Schwarzschild radius (mass M)	r_{0g}	$\alpha\,\pi^{-1}\,\lvert M \rvert_g$	$\alpha\,\pi^{-1}\,\lvert M \rvert_g\,\lambda_g$	$\lvert M \rvert_m \times 1.484 \times 10^{-27}\ \text{m}$
black-hole temperature (mass M)	T_g	$(8\,\pi\,\alpha\,\lvert M \rvert_g)^{-1}$	$(8\,\pi\,\alpha\,\lvert M \rvert_g)^{-1}\,k_g$	$\lvert M \rvert_m^{-1} \times 1.227 \times 10^{23}\ \text{K}$
black-hole density (mass M)	d_g	$\tfrac{3}{4}\,\pi^2\,\alpha^{-3}\,\lvert M \rvert_g^{-2}$	$\tfrac{3}{4}\,\pi^2\,\alpha^{-3}\,\lvert M \rvert_g^{-2}\, m_g \cdot \lambda_g^{-3}$	$\lvert M \rvert_m^2 \times 7.299 \times 10^{79}\ \text{kg·m}^{-3}$

Quantum force magnitudes of the electron and masstron

Force quanta of elementary particles	Code	Unit value	Quantum attribute-factor configuration		SI Value to 4 significant digits
Table VIII. Electron forces		**SE**			
electron inertial-force quantum	f_{ic}	137.036...	$\alpha^{-1} f_{ee} = \alpha^{-1} f_{me}$	(proportional to $m_e \cdot \lambda_e \cdot t_e^{-2}$)	3.374×10^{-2} N
electron electric-force quantum	f_{ee}	1	$\alpha f_{ic} = f_{ee}$	(proportional to $q_e^2 \cdot \lambda_e^{-2}$)	2.462×10^{-4} N
electron magnetic-force quantum	f_{me}	1	$\alpha f_{ic} = f_{me}$	(proportional to $q_e^2 \cdot t_e^{-2}$)	2.462×10^{-4} N
f_{ic}-to-f_{ee} force-magnitude ratio	α^{-1}		$f_{ic} f_{ee}^{-1}$		137.036...
f_{ic}-to-f_{me} force-magnitude ratio	α^{-1}		$f_{ic} f_{me}^{-1}$		137.036...
Table IX. Masstron forces		**SG**			
masstron inertial-force quantum	f_{ig}	137.036...	$\alpha^{-1} f_{gg}$	(proportional to $m_g \cdot \lambda_g \cdot t_g^{-2}$)	1.405×10^{41} N
masstron gravitational-force quantum	f_{gg}	1	$\alpha f_{ig} = f_{gg}$	(proportional to $m_g^2 \cdot \lambda_g^{-2}$)	1.025×10^{39} N
f_{im}-to-f_{gm} force-magnitude ratio	α^{-1}	137.036...	$f_{ig} f_{gg}^{-1}$		137.036...
Table X. Masstron-to-electron ratio					
SG-to-SE force ratio (inertial)	γ^2	4.166×10^{42}	$f_{ig} f_{ic}^{-1}$		4.166×10^{42}
SG-to-SE force ratio	γ^2	4.166×10^{42}	$f_{gg} f_{ee}^{-1}$ and $f_{gg} f_{me}^{-1}$		4.166×10^{42}

Quantum attributes of the electron bound in the Bohr hydrogen atom

Attribute name	Code (in Bohr units)	SE value (n = 1)	Quantum attribute-factor configuration	Historical formula (using constants)	SI value (n = 1)
Table XI. Bound electron quantum attributes of the Bohr hydrogen atom					
orbit (Bohr) radius	r_n	$(2\pi\,\alpha)^{-1}$	$(2\pi)^{-1}(n\,\alpha^{-1})(n\,\lambda_e)$	$2n^2h^2\epsilon_0(2\pi\,m_e\,q_e^2)^{-1}$	5.293×10^{-11} m
orbit length	λ_n	α^{-1}	$(n\,\alpha^{-1})(n\,\lambda_e)$	$2n^2h^2\epsilon_0(m_e\,q_e^2)^{-1}$	3.326×10^{-10} m
orbit period	$t_n = \omega_n^{-1}$	α^{-2}	$(n\,\alpha^{-1})^2(n\,t_e)$	$4n^3h^3\epsilon_0^2(m_e\,q_e^4)^{-1}$	1.520×10^{-16} s
orbit frequency	$\omega_n = t_n^{-1}$	α^2	$(n\,\alpha^{-1})^{-2}(n\,t_e)^{-1}$	$m_e\,q_e^4(4n^3h^3\epsilon_0^2)^{-1}$	6.577×10^{15} s^{-1}
orbit linear speed	$s_n = \lambda_n\,\omega_n$	α	$(n\,\alpha^{-1})^{-1}\lambda_e\cdot t_e^{-1}$	$q_e^2(2\,n\,h\,\epsilon_0)^{-1}$	2.187×10^6 m·s^{-1}
orbit linear momentum	$p_n = m_e\,s_n$	α	$m_e(n\,\alpha^{-1})^{-1}\lambda_e\cdot t_e^{-1}$	$m_e\,q_e^2(2\,n\,h\,\epsilon_0)^{-1}$	1.992×10^{-24} kg·m·s^{-1}
orbit angular momentum	$L_n = p_n\,r_n$	$(2\pi)^{-1}$	$m_e\,n\,(2\pi)^{-1}\lambda_e^2\,t_e^{-1}$	$n\,h\,(2\,\pi)^{-1}$	1.055×10^{-34} kg·m^2·s^{-1}
Table XII. Energy quantum attributes of the Bohr hydrogen atom					
Bohr energy	$E_n = +\,m_e\,\lambda_n^2\,t_n^{-2}$	$+\alpha^2$	$+n^{-2}\alpha^2\,E_e$	$+m_e\,q_e^4(2\,n\,h\,\epsilon_0)^{-2}$	$+4.358\times10^{-18}$ J
kinetic energy	$K_n = +\,\tfrac{1}{2}E_n$	$+\tfrac{1}{2}\,\alpha^2$	$+\tfrac{1}{2}\,n^{-2}\alpha^2\,E_e$	$+\tfrac{1}{2}\,m_e\,q_e^4(2\,n\,h\,\epsilon_0)^{-2}$	$+2.179\times10^{-18}$ J
potential energy	$U_n = -\,E_n$	$-\alpha^2$	$-n^{-2}\alpha^2\,E_e$	$-m_e\,q_e^4(2\,n\,h\,\epsilon_0)^{-2}$	-4.358×10^{-18} J
total energy	$E_T = K_n + U_n = -\,\tfrac{1}{2}E_n$	$-\tfrac{1}{2}\,\alpha^2$	$-\tfrac{1}{2}\,n^{-2}\alpha^2\,E_e$	$-\tfrac{1}{2}\,m_e\,q_e^4(2\,n\,h\,\epsilon_0)^{-2}$	-2.179×10^{-18} J

Quantum attributes of the electron bound in the Bohr hydrogen atom (concluded)

Attribute name	Code (in Bohr units)	SE value (n = 1)	Historical formula (using constants)	SI value (n = 1)

Table XIII. Photon emission quantum units of the Bohr hydrogen atom   Orbit numbers: i < j

Attribute name	Code (in Bohr units)	SE value (n = 1)	Historical formula (using constants)	SI value (n = 1)
frequency	$\nu_{ij} = \nu_R\,(i^{-2} - j^{-2})$	$\tfrac{1}{2}\,\alpha^2\,t_c^{-1}\,(i^{-2} - j^{-2})$	$m_e\,q_e^{\,4}\,(8\,h^3\,\epsilon_0^{\,2})^{-1}\,(I^{-2} - j^{-2})$	$3.289 \times 10^{15}\ s^{-1}\,(I^{-2} - j^{-2})$
(wave-length)$^{-1}$	$(\lambda_{ij})^{-1} = R_\infty\,(i^{-2} - j^{-2})$	$\tfrac{1}{2}\,\alpha^2\,\lambda_c^{-1}\,(i^{-2} - j^{-2})$	$m_e\,q_e^{\,4}\,(8\,c\,h^3\,\epsilon_0^{\,2})^{-1}\,(I^{-2} - j^{-2})$	$1.097 \times 10^7\ m^{-1}\,(I^{-2} - j^{-2})$
speed	$c = \lambda_{ij}\,\nu_{ij}$	$\lambda_c \cdot t_c^{-1}$	c	$2.998 \times 10^9\ m\cdot s^{-1}$

About the Author

The author in 1974

The author was born and raised in the Midwest during the height of the Great Depression and graduated from *Fairbanks High School* in Alaska. Fifty years ago, while a USAF pilot, he began pondering about aspects of inertial force. Then, over thirty years ago, his interest increased after attending a modern physics course while earning BSEE and MBA degrees at the *University of Washington* in Seattle. Thereafter, he became more and more interested in inertial force until reading about it and about other enigmas that pertain to our physical environment became his hobby. He spent countless hours in physics libraries and, over the decades, has purchased a great number of physics books. Now, he possesses an ever-growing library of them. He struggles through the more-professional books, not comprehending much of their arcane and highly-mathematical content, mainly in those thick books by Peebles, Wheeler, et alii. Nevertheless, he does obtain a good feeling for the limited range of information that he does seek.

Of course, as a hobby, this was not his day job. After leaving the USAF as a pilot in Normandy, he married there and helped his wife run her parents' restaurant, the finest on the highway midway between Paris and the resorts bordering upon the English Channel, with the finest international clientele. Years later, because a new super highway bypassed the restaurant and taxes paid by merchants increased, as France became more and more socialistic, he returned to the United States with his family.

Other work was as a camp manager at various remote construction sites in Africa and Alaska. Besides the restaurant in France, he was a French restaurateur and chef de cuisine in Florida and in Washington state. He was a professor of computer systems technology at *Memphis State University* in Memphis and a professor of computer science at *Western Washington University* in Bellingham near the Canadian border. Before retiring in San Diego, California, he was a quality analyst there for electronic document-storage and -retrieval systems and a technical writer creating, for those systems, user manuals in English and translating them into French.

www.ingramcontent.com/pod-product-compliance
Lightning Source LLC
Chambersburg PA
CBHW022023170526
45157CB00003B/1334